Potassium in Human Health

Edited by Jie Tang

Published in London, United Kingdom

Potassium in Human Health
http://dx.doi.org/10.5772/intechopen.94820
Edited by Jie Tang

Contributors
Shakuntala S. Patil, Sachin M. Patil, Elena Lastraioli, Jessica Iorio, Lisa Lastraioli, Jun Feng, Frank W. Sellke, Shawn Kant, Jie Tang, Sairah Sharif, Olive Tang

First published in London, United Kingdom, 2022 by IntechOpen
IntechOpen is the global imprint of INTECHOPEN LIMITED, registered in England and Wales, registration number: 11086078, 5 Princes Gate Court, London, SW7 2QJ, United Kingdom
Printed in Croatia

British Library Cataloguing-in-Publication Data
A catalogue record for this book is available from the British Library

Additional hard and PDF copies can be obtained from orders@intechopen.com

Potassium in Human Health
Edited by Jie Tang
p. cm.

This title is part of the Physiology Book Series, Volume 12
Topic: Human Physiology
Series Editor: Tomasz Brzozowski
Topic Editor: Kunihiro Sakuma
Associate Topic Editor: Kotomi Sakai

Print ISBN 978-1-78985-731-3
Online ISBN 978-1-78985-732-0
eBook (PDF) ISBN 978-1-83880-835-8
ISSN 2631-8261

For EU product safety concerns:
IN TECH d.o.o., Prolaz Marije Krucifikse Kozulić 3, 51000 Rijeka, Croatia,
info@intechopen.com or visit our website at intechopen.com.

IntechOpen Book Series
Physiology
Volume 12

Aims and Scope of the Series

Modern physiology requires a comprehensive understanding of the integration of tissues and organs throughout the mammalian body, including the cooperation between structure and function at the cellular and molecular levels governed by gene and protein expression. While a daunting task, learning is facilitated by identifying common and effective signaling pathways mediated by a variety of factors employed by nature to preserve and sustain homeostatic life. As a leading example, the cellular interaction between intracellular concentration of Ca+2 increases, and changes in plasma membrane potential is integral for coordinating blood flow, governing the exocytosis of neurotransmitters, and modulating gene expression and cell effector secretory functions. Furthermore, in this manner, understanding the systemic interaction between the cardiovascular and nervous systems has become more important than ever as human populations' life prolongation, aging and mechanisms of cellular oxidative signaling are utilised for sustaining life. Altogether, physiological research enables our identification of distinct and precise points of transition from health to the development of multimorbidity throughout the inevitable aging disorders (e.g., diabetes, hypertension, chronic kidney disease, heart failure, peptic ulcer, inflammatory bowel disease, age-related macular degeneration, cancer). With consideration of all organ systems (e.g., brain, heart, lung, gut, skeletal and smooth muscle, liver, pancreas, kidney, eye) and the interactions thereof, this Physiology Series will address the goals of resolving (1) Aging physiology and chronic disease progression (2) Examination of key cellular pathways as they relate to calcium, oxidative stress, and electrical signaling, and (3) how changes in plasma membrane produced by lipid peroxidation products can affect aging physiology, covering new research in the area of cell, human, plant and animal physiology.

Meet the Series Editor

Prof. Dr. Thomas Brzozowski works as a professor of Human Physiology and is currently a Chairman at the Department of Physiology and is V-Dean of the Medical Faculty at Jagiellonian University Medical College, Cracow, Poland. His primary area of interest is physiology and pathophysiology of the gastrointestinal (GI) tract, with a major focus on the mechanism of GI mucosal defense, protection, and ulcer healing. He was a postdoctoral NIH fellow at the University of California and the Gastroenterology VA Medical Center, Irvine, Long Beach, CA, USA, and at the Gastroenterology Clinics Erlangen-Nuremberg and Munster in Germany. He has published 290 original articles in some of the most prestigious scientific journals and seven book chapters on the pathophysiology of the GI tract, gastroprotection, ulcer healing, drug therapy of peptic ulcers, hormonal regulation of the gut, and inflammatory bowel disease.

Meet the Volume Editor

Jie Tang, MD, MPH, is an academic nephrologist and associate professor of Medicine at Albert Medical School, Brown University, USA. His research interest is in electrolyte disorders and bone mineral metabolism. Dr. Tang has served on journal editorial boards and published many articles in peer-reviewed journals. He is also a well-regarded clinician-educator, mentoring medical students, residents, and nephrology fellows. He gives lectures every year on national and international stages and has authored book chapters on various topics. He is a fellow of the American Society of Nephrology and an active member of the International Society of Nephrology. Dr. Tang is currently serving on the medical advisory boards for the National Kidney Foundation and End-Stage Renal Disease Network.

Contents

Preface

Potassium is the most abundant cation in the human body. It is an essential mineral that regulates intracellular enzyme function and helps determine neuromuscular and cardiovascular tissue excitability.

Adequate dietary intake is important to maintain body potassium store. In the United States, the average daily potassium intake is far from the recommended minimal target of 4700 mg per day. Population studies have demonstrated that higher dietary potassium intake helps reduce salt and water retention, mediate acid-base balance, protect against kidney stone formation and bone mineral loss, control blood pressure, and reduce the risk of type II diabetes mellitus (DM). It has also been associated with reduced risk of stroke, cardiovascular disease (CVD) events, and mortality.

Poor intake, intracellular shifting, or excessive loss can lead to hypokalemia, a potassium deficient state. Excessive potassium intake and/or extracellular shifting combined with reduced excretion will lead to hyperkalemia, a potassium excess state. Disturbances in potassium homeostasis are common and can result in life-threatening complications, including arrhythmia and cardiac arrest.

Managing potassium disorders is an essential part of daily clinical practice. But it often presents a huge challenge and can be intimidating at times for students, trainees, and even seasoned clinicians. It is crucial to have a systemic way of approaching potassium disorders. This approach should be based upon a solid understanding of the pathophysiology of potassium derangement.

This book is a concise, easy-to-read reference that provides clinicians and other healthcare providers with an in-depth understanding of the pathophysiology of potassium disorders in various clinical settings, as well as a systemic diagnostic and treatment approach based upon existing evidence. It is a useful companion book for students in the health professions, mid-level providers, house staff trainees, and clinicians responsible for the management of derangement in potassium homeostasis.

Jie Tang, MD., MPH, FASN
Division of Kidney Diseases and Hypertension,
Department of Medicine,
Alpert Medical School of Brown University,
Providence, Rhode Island, USA

Section 1

Introduction

Chapter 1

Introductory Chapter: Potassium in Human Health

Jie Tang and Olive Tang

1. Introduction

Potassium is an essential mineral in the maintenance of cellular integrity and physiologic homeostasis. The total body potassium store is about 45 millimoles per kilogram of body weight, most of which majority resides intracellularly. This introductory chapter will focus on dietary potassium and review the beneficial impacts of potassium on health.

2. Dietary potassium

Adequate dietary intake is important to maintain body potassium stores. Potassium is naturally present in a wide variety of foods, including fruits, vegetables, meats, and many common beverages, and can be found in nutritional supplements. Potassium content tends to be higher in fruits and vegetables compared to meats and grains. **Table 1** lists the potassium content in common foods.

High (>200 mg/serving)	**Moderate (100–200 mg/serving)**	**Low (<100 mg/serving)**
Fruits		
Raisin (1 cup = 1086 mg)	Pear (1 = 198 mg)	Grapes (10 = 93 mg)
Pineapple (1 = 986 mg)	Apple, medium (1 = 195 mg)	Lemon (1 = 58 mg)
Papaya (1 = 781 mg)	Peach, medium (1 = 193 mg)	
Mango (1 = 564 mg)	Raspberry, fresh (1 cup = 186 mg)	
Kiwi, sliced (1 cup = 562 mg)	Watermelon, diced (1 cup = 170 mg)	
Banana (1 = 422 mg)	Tangerines, medium (1 = 146 mg)	
Cantaloupe, diced (1cup = 417 mg)	Plums, fresh (1 = 104 mg)	
Honeydew, diced (1 cup = 388 mg)	Blueberries (1 cup = 112 mg)	
Grapefruit (1 = 350 mg)		
Orange, medium (1 = 237 mg)		
Blackberries (1 cup = 233 mg)		
Strawberry, whole (1 cup = 220 mg)		
Vegetables		
Avocado, medium (1 = 975 mg)	Carrots, fresh (1/2 cup = 177 mg)	Cucumber slices (1/2 cup = 88 mg)
Potato, medium (1 = 897 mg)	Onion, medium (1 = 161 mg)	Lettuce (1 cup = 87 mg)

IntechOpen

High (>200 mg/serving)	Moderate (100–200 mg/serving)	Low (<100 mg/serving)
Spinach (1 cup = 840 mg)	Cabbage, chopped (1 cup = 151 mg)	Radish, medium (1 = 10 mg)
Sweet potato (1 = 438 mg)	Broccoli, chopped (1/2 cup = 147 mg)	
Winter squash (1 cup = 406 mg)	Peas, whole (1 cup = 126 mg)	
Corn, medium ear (1 = 368 mg)	Eggplant (1 cup = 122 mg)	
Brussels sprouts (1 cup = 342 mg)	Lettuce leaf (1 cup = 109 mg)	
Celery, raw (1 cup = 312 mg)		
Cauliflower, fresh (1 cup = 304 mg)		
Tomato, fresh (1 = 290 mg)		
Asparagus (1 cup = 271 mg)		
Carrots, fresh (1 = 230 mg)		
Mushroom, raw (1 cup = 220 mg)		
Green beans, fresh (1 cup = 209 mg)		
Animal products		
Clams (3 oz. = 534 mg)	Ham patties, cooked (1 = 146 mg)	Egg, raw, large (1 = 69 mg)
Cod (3 oz. = 449 mg)		
Salmon, raw (3 oz. = 309 mg)		
Turkey breast (100 gm = 275 mg)		
Beef, ground, 85% lean (3 oz. = 270 mg)		
Shrimp, cooked (100 gm = 259 mg)		
Lean beef, cooked (3 oz. = 224 mg)		
Chicken (100 gm = 223 mg)		
Blue crabs, cooked (3 oz. = 220 mg)		
Quail cooked (100 gm = 216 mg)		
Beverages/diary products		
Prune juice (8 FL oz. = 707 mg)	Wine (5 FL oz. = 147 mg)	Beer (12 FL oz. = 96 mg)
Orange juice (8 FL oz. = 596 mg)	Chocolate ice cream (1/2 cup = 164 mg)	American cheese (1 oz. = 79 mg)
Ensure (8 FL oz. = 441 mg)	Vanilla ice cream (1/2 cup = 131 mg)	Coffee (6 FL oz. = 52 mg)
Yogurt, fruit (8 FL oz. = 475 mg)		Cranberry juice (8 FL oz. = 46 mg)
Milk whole (8 FL oz. = 322 mg)		Cheddar cheese (1 oz. = 28 mg)
Soy milk (8 FL oz. = 287 mg)		Cola (12 FL oz. = 4 mg)
Others		
Breakfast cereal (1 cup = 293 mg)	Energy bar (1 = 179 mg)	Granola bar (1 = 94 mg)
White chocolate (3 oz. = 243 mg)	Chocolate, dark (1 oz. = 158 mg)	Rice, white, cooked (1 cup = 55 mg)
	Wheat flour (1 cup = 134 mg)	Butter (1 tbsp = 3 mg)

Source: http://www.nal.usda.gov.

Table 1.
Dietary potassium content in common foods.

A typical western diet includes more grains and meat and fewer fruits and vegetables, corresponding to a diet lower in potassium and higher in sodium. According to the National Health and Nutrition Examination Survey in 2009–2010, the average daily potassium intake in U.S. adults was only 2650 mg [1], far from the recommended minimal target of 4700 mg per day. This large deficit in potassium intake at the population level led to potassium being identified as a shortfall nutrient by the Dietary Guidelines for Americans 2010 Advisory Committee [2].

In general, higher dietary potassium intake is associated with many health benefits, including better blood pressure and blood glucose control [3]. However, in cases of advanced kidney diseases, dietary potassium intake may need to be restricted to avoid hyperkalemia [4].

3. Effect of potassium on health

In this chapter, we will review the existing evidence on the effect of potassium on health. Most published studies show the health benefits of higher potassium intake either by consumption of more fruits and vegetables, salt substitutes, or potassium supplementation.

3.1 Cell structure and function

Potassium is an important intracellular ion critical in the maintenance of cellular osmolarity and function. The resting cell membrane potential is a crucial component of cellular function especially for excitable cells, such as muscles and nerves. It is created and maintained by the movement of sodium and potassium through corresponding channels and transporters in the cell membrane, specifically, the Na-K-ATPase and K-selective outward ion channels. Disturbances in extracellular potassium concentration can affect membrane potential and alter cellular function [5].

Potassium is also important for several intracellular processes including ribosome structure and function. Rozov et al. (2019) examined the role of potassium in ribosomal protein synthesis at the three-dimensional level using long-wavelength X-ray crystallography [6]. They found that potassium stabilized main functional ligands such as messenger RNA and transfer RNAs, as well as ribosomal RNAs and ribosomal proteins, through direct interactions with nitrogen and oxygen atoms of nucleotide bases and polypeptide or sugar-phosphate residues. Potassium helped preserve ribosome structural and functional integrities, by coordinating the conformational rearrangements in the decoding center upon binding of aminoacyl-tRNA, allowing tighter binding of mRNA.

In the mitochondria, several potassium channels have been described in the inner mitochondrial membrane: ATP-regulated potassium channel, Ca^{2+}-activated potassium channel, the voltage-gated Kv1.3 potassium channel, and the two-pore domain TASK-3 potassium channel [7]. They help preserve mitochondrial membrane potential, oxidative phosphorylation coupling, and matrix volume, and regulate several key mitochondrial metabolic actions [8, 9].

3.2 Blood pressure (BP)

Hypertension is a leading cause of heart attacks, strokes, and other end-organ damage. The antihypertensive effect of increased potassium intake is likely multifactorial. Blood pressure is in part determined by body salt and fluid status; potassium is an essential ion for salt and fluid volume regulation. Potassium intake can modulate the activity of sodium-chloride cotransporter (NCC), a key channel for renal salt reabsorption located at distal nephron, and a special potassium channel, Kir4.1, located in the basolateral side of distal convoluted tubule [10]. Kir4.1 plays a key role in sensing plasma potassium. Low potassium intake (LKI) activates Kir4.1, hyperpolarizes cells at the distal convoluted tubule which, in turn, increases the abundance of NCC channel at the luminal side and ultimately leads to salt retention and higher blood pressure [10]. Increased potassium intake increases urinary salt excretion, thereby reducing blood pressure. In addition, higher potassium intake can directly lower the blood pressure by vasodilation due to vascular endothelium-dependent smooth muscle cell hyperpolarization [11]. Finally, potassium intake can also modulate baroreceptor sensitivity, and reduce vascular sensitivity to catecholamines [12].

Large epidemiologic studies have consistently appear significant but inverse associations between potassium intake and BP, independent of dietary salt intake [13–15]. Controlled interventional trials of vegetarian diets (typically higher in potassium content) in both normotensive and hypertensive populations also demonstrated a BP-lowering effect [16, 17]. According to the seminal Dietary Approaches to Stop Hypertension (DASH) trial, a diet rich in fruits, vegetables, and reduced saturated and total fat (potassium content: 2776 vs. 1447 mg/day in the control group) substantially lowered blood pressure [18]. Unfortunately, prospective clinical trials examining the independent BP effect from oral potassium supplementation have yielded conflicting results [19–21]. The effect size from potassium intake alone may be small; therefore, a larger number of study participants and a longer duration of follow-up are needed. To address these concerns, several meta-analyses were conducted and again confirmed the BP reduction effect from potassium supplementation [22–24]. A meta-analysis of 33 randomized controlled trials involving 2609 participants in whom potassium supplementation was the only difference between the intervention and control conditions, showed that potassium supplementation was associated with a significant reduction in mean (95% confidence interval) systolic BP (SBP) and diastolic BP (DBP) of −3.11 mm Hg (−1.91 to −4.31 mm Hg) and – 1.97 mm Hg (−0.52 to −3.42 mm Hg), respectively [22]. However, a small meta-analysis (n = 483) with a short duration of follow-up failed to show statistically significant BP reduction by potassium supplementation. Despite the lack of statistical significance in this meta-analysis, both SBP and DBP were reduced with potassium supplementation (mean difference in SBP: −11.2, 95% CI: −25.2 to 2.7; mean difference in DBP: −5.0, 95% CI: −12.5 to 2.4) [25].

The BP-lowering effect by potassium appears to be modified by race, with individuals who identified as being Black, appearing to be particularly sensitive to the BP-lowering effect of potassium. According to a randomized, double-blind clinical trial, the reduction in SBP reached a mean of 19 mmHg compared to the 3.4 mmHg reduction in the overall study population [26]. Since there is a physiologic interdependency of sodium and potassium, the BP-lowering effect of potassium also appears to be modified by salt intake. Clinical studies have demonstrated that dietary potassium intake did not affect BP after salt restriction [21, 27]. However, higher potassium intake blunted the rising in BP after salt loading [28]. These were consistent with the epidemiologic studies showing that the BP-lowering effect of potassium was more discernible in participants who had a high sodium intake [22, 29]. As a result, a dietary sodium-to-potassium ratio is now being accepted as a better metric for cardiovascular risk as compared to the separate dietary sodium or potassium values alone [30]. A 1:1 molar ratio of sodium to potassium is considered beneficial for health [31].

3.3 Insulin resistance and diabetes

Diabetes is a global epidemic affecting over 34 million Americans and 422 million people worldwide [32]. Studies examining the association between dietary potassium intake and diabetes risk have been mixed and mostly unrevealing. This could be due to the lack of standardized measures of dietary potassium intake, and possibly a small effect size from dietary potassium manipulation alone. Total body potassium depletion has been linked to an increased risk for insulin resistance and diabetes. Medically-induced hypokalemia has been associated with reduced pancreatic β-cell sensitivity and insulin release in response to hyperglycemia. Potassium supplementation corrected the defective insulin response to glucose, further implicating hypokalemia as the direct cause of glucose intolerance [33, 34]. Adding to the existing evidence, observational studies showed a significant inverse association between serum potassium levels and risk of incident diabetes [35, 36]. Genetic studies also demonstrated that mutations in genes coding for potassium channels on pancreatic β-cells affect insulin secretion. The *KCNJ11* gene encodes for the ATP-sensitive potassium channel in the pancreatic beta-cell and mutations in this gene have been linked to neonatal diabetes [37], with single nucleotide polymorphisms associated with increased susceptibility to type II diabetes [38]. Another potassium channel gene, *KCNQ1*, is also involved in insulin secretion and its polymorphisms have been associated with an increased risk for type II diabetes in Asian women [39].

3.4 Cardiovascular disease and survival

Physiologic increases in potassium concentration can inhibit superoxide anion formation by endothelial cells and monocytes/macrophages, leading to a reduction in oxidative stress and inflammation [40]. Higher physiological potassium concentrations can reduce vascular smooth muscle cell proliferation and inhibit vascular thrombosis, leading to a reduction in ischemic events [41]. Oral potassium supplementation can also significantly improve endothelial function as measured by brachial artery flow-mediated dilatation, increase arterial compliance as assessed by carotid-femoral pulse wave velocity, decrease left ventricular mass, and improve left ventricular diastolic function [42]. More importantly, since hypertension and diabetes are the two strongest risk factors for cardiovascular disease (CVD) and CVD mortality, higher potassium intake can theoretically reduce the risk of CVD and improve survival by lowering BP and improving insulin sensitivity. Indeed, the protective effect of potassium on CVD and survival is supported by a large randomized controlled trial of 1981 Chinese veterans, in whom the use of potassium-enriched salt led to a significant reduction in CVD mortality (age-adjusted hazard ratio: 0.59; 95% CI: 0.37–0.95) [43]. A meta-analysis of 11 studies that included 247,510 adult participants (follow-up 5–19 years) showed that higher potassium intake (by 1.64 g per day), was associated with a 21% lower risk of stroke (risk ratio (RR): 0.79; 95% CI: 0.68–0.90; $p = 0.0007$), and trended toward lower risk of CVD including coronary heart disease, which was statistically significant after the exclusion of a single cohort, based on sensitivity analysis (RR: 0.74; 95% CI: 0.60–0.91; $p = 0.0037$) [44]. In this meta-analysis, potassium intake was assessed either by 24-h dietary recall, food frequency questionnaire, or 24-h urinary excretion. Another meta-analysis of nine cohort studies also detected a protective effect of higher potassium intake on the risk of incident stroke (RR 0.76, 95% CI: 0.66–0.89). However, potassium intake had a non-significant relation with incident cardiovascular disease (RR 0.88, 95% CI: 0.70–1.10) and coronary heart disease (RR 0.96, 95% CI: 0.78–1.19). Mortality analyses were not performed in this meta-analysis due to missing data [45].

3.5 Acid-base regulation

Tight blood pH control within a narrow range is essential for proper cellular and organ function.

The kidney plays a key role in bicarbonate reabsorption and generation, as well as proton excretion, to maintain acid-base homeostasis. This is achieved by ammonia-genesis in the proximal tubule, where glutamine is transported internally and metabolized into ammonia, bicarbonate, and glucose [46]. This process can be dramatically upregulated in response to metabolic acidosis.

In animals under dietary potassium deprivation, protein expression of the glutamine transporter (system N transporter, SN1) in the proximal tubule was stimulated, along with an upregulation of key enzymes in glutamine catabolism including glutaminase, glutamate dehydrogenase, and phosphoenolpyruvate carboxykinase, leading to an enhanced ammoniagenesis. The increase in urinary ammonia formation and excretion is observed within 2 days of potassium deprivation while the blood potassium level is still maintained within normal limits. The observed effect was independent of systemic metabolic acidosis or changes in volume status [47]. In a human study where participants had moderate potassium depletion acutely induced by dietary potassium restriction (with sodium substitution), a significant increase in urine ammonia excretion was observed along with a rise in plasma bicarbonate and a reduction in blood acidity. None of the participants had increases in mineralocorticoid levels or evidence of chloride depletion [48]. In another study of human volunteers, acute potassium depletion was induced by dietary restriction alone or with additional use of potassium binding resin. There were increases in urinary excretions of both net acid and ammonium [49]. Similar findings were observed in patients with chronic potassium depletion. In a study of eight patients with chronic potassium depletion and 20 healthy controls, potassium depletion was associated with an almost two-fold higher ammoniagenesis measured by the amount of ammonia excreted in the urine plus the amount added to the venous blood. There was also an increased urinary ammonia excretion and an enhanced renal extraction of glutamine (56.6 ± 5.9 mumol/min/1.73 m^2 vs. 34.6 ± 3.1 in controls). Total ammonia production was inversely correlated with serum potassium and directly correlated with urine flow. Stepwise multiple regression analysis again showed that serum potassium was the predominant factor affecting renal ammonia production [50].

While hypokalemia stimulates ammoniagenesis, hyperkalemia suppresses renal ammonia production in the proximal tubule and potentiates the development of hyperchloremic metabolic acidosis. In an experiment using renal tissue from animals, treatment with a higher potassium bath (7–25 vs. 4–5 meq/L) resulted in significantly less ammonia production and suppression of the catabolic pathway of glutamine [51]. In an animal study, chronic potassium loading leading to hyperkalemia resulted in a 40% reduction in urinary ammonium excretion and a 50% reduction in whole kidney ammonium production. These animals also developed metabolic acidosis [52]. Similar findings were observed in a genetic animal model of hyperkalemia (DCT-CA-SPAK mice), who developed co-existing metabolic acidosis and reduced ammonia excretion [53]. These mice were found to have reduced proximal tubule expressions of key enzymes in ammoniagenesis and reduced ammonia transporter expression in the distal nephron. Treatment with hydrochlorothiazide corrected the hyperkalemia and metabolic acidosis, increased ammonia excretion, and normalized both enzyme levels in ammoniagenesis and distal nephron ammonia transporter expression [53].

3.6 Kidney stones

Kidney stone disease is common in the general population with an estimated prevalence of around 10–15% in males and 3–5% in females [54]. Calcium-based kidney stones are the most common type (>80%), with high urine calcium excretion being a strong risk factor for stone formation [55]. Potassium is a key regulator of urinary calcium excretion at the thick ascending limb of the loop of Henle. The Na/K/2Cl cotransporter (NKCC) reabsorbs sodium and potassium with potassium returning to the lumen via the renal outer medullary K channel (ROMK). This, in turn, results in a more positively charged milieu that promotes the reabsorptions of calcium through paracellular routes. Loss-of-function mutations in *KCNJ1* coding for ROMK results in Bartter syndrome, with increased urinary calcium loss [56]. Body stores of potassium can affect potassium intraluminal availability, and therefore play a role in this intricate regulation of calcium excretion. Furthermore, potassium deficiency has been shown to increase phosphorus excretion, which may lead to an increase in vitamin D and subsequent modulation of calcium homeostasis [57]. According to a study in adult volunteers, potassium administration increases, and potassium deprivation reduces urinary calcium excretion [58]. In this study, dietary potassium deprivation with a mean intake of 67 ± 8 mmol/day led to a significant increase in daily urinary Ca excretion (+1.31 ± 0.25 mmol/day, $p < 0.005$), which was later normalized after the administrations of either potassium chloride or potassium bicarbonate (90 mmol/day). An independent association between dietary potassium intake and incident kidney risk was demonstrated in a large epidemiologic study [59]. Taylor et al. (2016) prospectively examined potassium intake and risk of incident kidney stones in the Health Professionals Follow-Up Study (n = 42,919), the Nurses' Health Study I (n = 60,128), and the Nurses' Health Study II (n = 90,629). After multivariable adjustment, there was a strong inverse association between potassium intake and risk of incident stones with a HR of 0.44 (95% CI: 0.36–0.53) for the HPFS, 0.57 (95% CI: 0.45–0.72) for the NHS I, and 0.67 (95% CI: 0.57–0.78) for the NHS II (all P values for trend <0.001) [59].

3.7 Bone health

Bone health is determined by body calcium stores and acid-base balance with both calcium depletion and acidosis being among the strongest risk factors for bone mineral loss [60]. As mentioned earlier, potassium is a key regulator of both calcium homeostasis and acid-base balance. Potassium depletion can lead to renal calcium loss and stimulation of ammoniagenesis whereas potassium supplementation will preserve body calcium storage and may inhibit ammoniagenesis, if hyperkalemia is present. Therefore, potassium has long been thought to play an important role in bone health and osteoporosis prevention. The DASH diet, typically high in potassium, was also found to associate with improved biomarker profile for bone turnover [61]. However, epidemiologic studies using fruits and vegetables as a marker of potassium intake have reported conflicting results [62–64]. This could be due to differences in various other dietary nutrients (such as acid, protein, calcium, phosphorus, and vitamins), which may be confounding the association between potassium and bone turnover. Indeed, findings from studies focusing on specific dietary potassium have been rather consistent [65–68]. In an analysis of the Framingham Heart Study, higher potassium intake was significantly associated with greater bone mineral density (BMD) and less decline in BMD in both men and women [65]. According to a large Korean population survey, dietary potassium intake was associated with improved bone mineral density in both older men (age > 50) and postmenopausal women [66]. Overall, participants in the highest tertile of potassium intake had a significantly higher total hip and femur neck BMD as compared to those in the lower tertile groups ($p = 0.014$, and 0.012 for total hip and femur neck respectively). Among postmenopausal women, those in the highest tertile of potassium intake also had significantly higher lumbar, total hip, and femur neck BMD as compared to those in lower potassium intake tertile groups ($p = 0.029$, 0.002, and 0.002 for lumbar spine, total hip, and femur neck respectively). Currently, there are only a few small clinical trials that have used potassium supplements to examine their effect on bone health. In a trial of 201 healthy older adults (age ≥ 65), daily supplementation with 60 mmol potassium citrate resulted in significantly increased bone mineral density at the lumbar spine and improved bone microarchitecture compared to placebo [69]. Similar findings were observed in another randomized, placebo-controlled trial of 233 older adults aged 50 years and older. Supplemental potassium bicarbonate at 1–1.5 mmol/kg daily for 3 months led to a significantly reduced biochemical markers of bone turnover and urinary calcium excretion [70]. However, an earlier randomized controlled trial of 276 postmenopausal women (age 55–65), supplementation with potassium citrate at either 18.5 or 55.5 mEq/day for 2 years did not reduce bone turnover or increase BMD at the hip or lumbar spine compared with placebo [71]. The differences in the study population may be the reason for the different study findings. Thus far, there is no conclusive evidence proving an independent effect of potassium intake on bone health and turnover. Future large-scale randomized trials are needed.

4. Conclusion

Potassium is an essential mineral important for maintaining cell and organ function. It is naturally present in a wide variety of foods. Adequate potassium intake can provide many health benefits, including better blood pressure control, lower risk for diabetes and cardiovascular disease, and improved overall survival.

Author details

Jie Tang[1*] and Olive Tang[2]

1 Division of Kidney Diseases and Hypertension, Department of Medicine, Alpert Medical School of Brown University, Providence, Rhode Island, USA

2 Johns Hopkins University School of Medicine, Baltimore, MD, USA

*Address all correspondence to: jie.tang@lifespan.org

References

[1] USDA. Potassium Intake of the U.S. Population. What We Eat in America, NHANES 2009-2010. USDA; 2012. https://www.ars.usda.gov/ARSUserFiles/80400530/pdf/DBrief/10_potassium_intake_0910.pdf

[2] DeSalvo KB, Olson R, Casavale KO. Dietary guidelines for Americans. JAMA. 2016;**315**(5):457-458

[3] Aaron KJ, Sanders PW. Role of dietary salt and potassium intake in cardiovascular health and disease: A review of the evidence. Mayo Clinic Proceedings. 2013;**88**(9):987-995

[4] Tang J. Conn's Current Therapy—Hypokalemia and Hyperkalemia. Elsevier; 2020. p. 338-341. ISBN: 9780323790062

[5] Stambrook PJ, Sachs HG, Ebert JD. The effect of potassium on the cell membrane potential and the passage of synchronized cells through the cell cycle. Journal of Cellular Physiology. 1975; **85**(2 Pt 1):283-291

[6] Rozov A et al. Importance of potassium ions for ribosome structure and function revealed by long-wavelength X-ray diffraction. Nature Communications. 2019;**10**(1):2519

[7] Laskowski M et al. What do we not know about mitochondrial potassium channels? Biochimica et Biophysica Acta. 2016;**1857**(8):1247-1257

[8] Szabo I, Zoratti M. Mitochondrial channels: ion fluxes and more. Physiological Reviews. 2014;**94**(2): 519-608

[9] Gomez-Puyou A et al. On the role of K+ on oxidative phosphorylation. The Journal of Biological Chemistry. 1970;**245**(20):5239-5247

[10] Wang MX et al. Potassium intake modulates the thiazide-sensitive sodium-chloride cotransporter (NCC) activity via the Kir4.1 potassium channel. Kidney International. 2018;**93**(4):893-902

[11] Haddy FJ, Vanhoutte PM, Feletou M. Role of potassium in regulating blood flow and blood pressure. American Journal of Physiology. Regulatory, Integrative and Comparative Physiology. 2006;**290**(3):R546-R552

[12] Skrabal F et al. Effect of moderate salt restriction and high potassium intake on pressor hormones, response to noradrenaline and baroreceptor function in man. Clinical Science. 1980;**59** (Suppl 6):157s-160s

[13] Intersalt: An international study of electrolyte excretion and blood pressure. Results for 24 hour urinary sodium and potassium excretion. Intersalt Cooperative Research Group. BMJ. 1988;**297**(6644):319-328

[14] Khaw KT, Rose G. Population study of blood pressure and associated factors in St Lucia, West Indies. International Journal of Epidemiology. 1982;**11**(4): 372-377

[15] Stamler J et al. Findings of the International Cooperative INTERSALT Study. Hypertension. 1991;**17**(1 Suppl): I9-I15

[16] Rouse IL et al. Blood-pressure-lowering effect of a vegetarian diet: Controlled trial in normotensive subjects. Lancet. 1983;**1**(8314-5):5-10

[17] Margetts BM et al. Vegetarian diet in mild hypertension: A randomised

controlled trial. British Medical Journal. 1986;**293**(6560):1468-1471

[18] Appel LJ et al. A clinical trial of the effects of dietary patterns on blood pressure. DASH Collaborative Research Group. The New England Journal of Medicine. 1997;**336**(16):1117-1124

[19] Poorolajal J et al. Oral potassium supplementation for management of essential hypertension: A meta-analysis of randomized controlled trials. PLoS One. 2017;**12**(4):e0174967

[20] Berry SE et al. Increased potassium intake from fruit and vegetables or supplements does not lower blood pressure or improve vascular function in UK men and women with early hypertension: A randomised controlled trial. The British Journal of Nutrition. 2010;**104**(12):1839-1847

[21] Grimm RH Jr et al. The influence of oral potassium chloride on blood pressure in hypertensive men on a low-sodium diet. The New England Journal of Medicine. 1990;**322**(9): 569-574

[22] Whelton PK et al. Effects of oral potassium on blood pressure. Meta-analysis of randomized controlled clinical trials. JAMA. 1997;**277**(20): 1624-1632

[23] Cappuccio FP, MacGregor GA. Does potassium supplementation lower blood pressure? A meta-analysis of published trials. Journal of Hypertension. 1991;**9**(5):465-473

[24] Geleijnse JM, Kok FJ, Grobbee DE. Blood pressure response to changes in sodium and potassium intake: A metaregression analysis of randomised trials. Journal of Human Hypertension. 2003;**17**(7):471-480

[25] Dickinson HO et al. Potassium supplementation for the management of primary hypertension in adults. Cochrane Database of Systematic Reviews. 2006;**3**:CD004641

[26] Svetkey LP et al. Double-blind, placebo-controlled trial of potassium chloride in the treatment of mild hypertension. Hypertension. 1987;**9**(5):444-450

[27] Smith SJ et al. Moderate potassium chloride supplementation in essential hypertension: Is it additive to moderate sodium restriction? British Medical Journal. 1985;**290**(6462):110-113

[28] Krishna GG. Effect of potassium intake on blood pressure. Journal of the American Society of Nephrology. 1990;**1**(1):43-52

[29] Filippini T et al. The effect of potassium supplementation on blood pressure in hypertensive subjects: A systematic review and meta-analysis. International Journal of Cardiology. 2017;**230**:127-135

[30] Iwahori T, Miura K, Ueshima H. Time to consider use of the sodium-to-potassium ratio for practical sodium reduction and potassium increase. Nutrients. Nutrients. 2017 Jul 5;**9**(7):700.

[31] World Health Organization. Diet, Nutrition and the Prevention of Chronic Disease: Report of a Joint WHO/FAO Expert Consultation. WHO; 2003

[32] WHO Fact sheets—Diabetes. 2021

[33] Gorden P. Glucose intolerance with hypokalemia. Failure of short-term potassium depletion in normal subjects to reproduce the glucose and insulin abnormalities of clinical hypokalemia. Diabetes. 1973;**22**(7):544-551

[34] Helderman JH et al. Prevention of the glucose intolerance of thiazide diuretics by maintenance of body potassium. Diabetes. 1983;**32**(2): 106-111

[35] Chatterjee R et al. Serum and dietary potassium and risk of incident type 2 diabetes mellitus: The Atherosclerosis Risk in Communities (ARIC) study. Archives of Internal Medicine. 2010;**170**(19):1745-1751

[36] Heianza Y et al. Low serum potassium levels and risk of type 2 diabetes: the Toranomon Hospital Health Management Center Study 1 (TOPICS 1). Diabetologia. 2011;**54**(4):762-766

[37] Abujbara MA et al. Permanent neonatal diabetes mellitus in Jordan. Journal of Pediatric Endocrinology & Metabolism. 2014;**27**(9-10):879-883

[38] Haghvirdizadeh P et al. KCNJ11: Genetic polymorphisms and risk of diabetes mellitus. Journal Diabetes Research. 2015;**2015**:908152

[39] Shin HD et al. Association of KCNQ1 polymorphisms with the gestational diabetes mellitus in Korean women. The Journal of Clinical Endocrinology and Metabolism. 2010;**95**(1):445-449

[40] McCabe RD et al. Potassium inhibits free radical formation. Hypertension. 1994;**24**(1):77-82

[41] Young DB, Ma G. Vascular protective effects of potassium. Seminars in Nephrology. 1999;**19**(5):477-486

[42] He FJ et al. Effects of potassium chloride and potassium bicarbonate on endothelial function, cardiovascular risk factors, and bone turnover in mild hypertensives. Hypertension. 2010;**55**(3):681-688

[43] Chang HY et al. Effect of potassium-enriched salt on cardiovascular mortality and medical expenses of elderly men. The American Journal of Clinical Nutrition. 2006;**83**(6):1289-1296

[44] D'Elia L et al. Potassium intake, stroke, and cardiovascular disease a meta-analysis of prospective studies. Journal of the American College of Cardiology. 2011;**57**(10):1210-1219

[45] Aburto NJ et al. Effect of increased potassium intake on cardiovascular risk factors and disease: Systematic review and meta-analyses. BMJ. 2013;**346**:f1378

[46] Rajkumar P, Pluznick JL. Acid-base regulation in the renal proximal tubules: Using novel pH sensors to maintain homeostasis. American Journal of Physiology. Renal Physiology. 2018;**315**(5):F1187-F1190

[47] Abu Hossain S et al. Cellular and molecular basis of increased ammoniagenesis in potassium deprivation. American Journal of Physiology. Renal Physiology. 2011;**301**(5):F969-F978

[48] Jones JW et al. Systemic and renal acid-base effects of chronic dietary potassium depletion in humans. Kidney International. 1982;**21**(2):402-410

[49] Tannen RL. The effect of uncomplicated potassium depletion on urine acidification. The Journal of Clinical Investigation. 1970;**49**(4): 813-827

[50] Tizianello A et al. Renal ammoniagenesis in humans with chronic potassium depletion. Kidney International. 1991;**40**(4):772-778

[51] Sleeper RS et al. Effects of in vitro potassium on ammoniagenesis in rat and

canine kidney tissue. Kidney International. 1982;**21**(2):345-353

[52] DuBose TD Jr, Good DW. Effects of chronic hyperkalemia on renal production and proximal tubule transport of ammonium in rats. The American Journal of Physiology. 1991;**260**(5 Pt 2):F680-F687

[53] Harris AN et al. Mechanism of hyperkalemia-induced metabolic acidosis. Journal of the American Society of Nephrology. 2018;**29**(5): 1411-1425

[54] Stamatelou KK et al. Time trends in reported prevalence of kidney stones in the United States: 1976-1994. Kidney International. 2003;**63**(5):1817-1823

[55] Tang J, Chonchol MB. Vitamin D and kidney stone disease. Current Opinion in Nephrology and Hypertension. 2013;**22**(4):383-389

[56] Simon DB et al. Genetic heterogeneity of Bartter's syndrome revealed by mutations in the K+ channel, ROMK. Nature Genetics. 1996;**14**(2): 152-156

[57] Breusegem SY et al. Differential regulation of the renal sodium-phosphate cotransporters NaPi-IIa, NaPi-IIc, and PiT-2 in dietary potassium deficiency. American Journal of Physiology. Renal Physiology. 2009;**297**(2):F350-F361

[58] Lemann J Jr et al. Potassium administration reduces and potassium deprivation increases urinary calcium excretion in healthy adults [corrected]. Kidney International. 1991;**39**(5): 973-983

[59] Ferraro PM et al. Dietary protein and potassium, diet-dependent net acid load, and risk of incident kidney stones. Clinical Journal of the American Society of Nephrology. 2016;**11**(10):1834-1844

[60] Tylavsky FA, Spence LA, Harkness L. The importance of calcium, potassium, and acid-base homeostasis in bone health and osteoporosis prevention. The Journal of Nutrition. 2008;**138**(1): 164S-165S

[61] Lin PH et al. The DASH diet and sodium reduction improve markers of bone turnover and calcium metabolism in adults. The Journal of Nutrition. 2003;**133**(10):3130-3136

[62] Lloyd T et al. Urinary hormonal concentrations and spinal bone densities of premenopausal vegetarian and nonvegetarian women. The American Journal of Clinical Nutrition. 1991;**54**(6): 1005-1010

[63] Ho-Pham LT, Nguyen ND, Nguyen TV. Effect of vegetarian diets on bone mineral density: A Bayesian meta-analysis. The American Journal of Clinical Nutrition. 2009;**90**(4):943-950

[64] Knurick JR et al. Comparison of correlates of bone mineral density in individuals adhering to lacto-ovo, vegan, or omnivore diets: A cross-sectional investigation. Nutrients. 2015;**7**(5): 3416-3426

[65] Tucker KL et al. Potassium, magnesium, and fruit and vegetable intakes are associated with greater bone mineral density in elderly men and women. The American Journal of Clinical Nutrition. 1999;**69**(4):727-736

[66] Kong SH et al. Dietary potassium intake is beneficial to bone health in a low calcium intake population: The Korean National Health and Nutrition Examination Survey (KNHANES) (2008-2011). Osteoporosis International. 2017;**28**(5):1577-1585

[67] Zhu K, Devine A, Prince RL. The effects of high potassium consumption on bone mineral density in a prospective cohort study of elderly postmenopausal women. Osteoporosis International. 2009;**20**(2):335-340

[68] New SA et al. Nutritional influences on bone mineral density: A cross-sectional study in premenopausal women. The American Journal of Clinical Nutrition. 1997;**65**(6):1831-1839

[69] Jehle S, Hulter HN, Krapf R. Effect of potassium citrate on bone density, microarchitecture, and fracture risk in healthy older adults without osteoporosis: A randomized controlled trial. The Journal of Clinical Endocrinology and Metabolism. 2013;**98**(1):207-217

[70] Dawson-Hughes B et al. Potassium bicarbonate supplementation lowers bone turnover and calcium excretion in older men and women: A randomized dose-finding trial. Journal of Bone and Mineral Research. 2015;**30**(11): 2103-2111

[71] Macdonald HM et al. Effect of potassium citrate supplementation or increased fruit and vegetable intake on bone metabolism in healthy postmenopausal women: A randomized controlled trial. The American Journal of Clinical Nutrition. 2008;**88**(2): 465-474

Section 2

Physiology and Pathophysiology

Chapter 2

Potassium Homeostasis

Shakuntala S. Patil and Sachin M. Patil

Abstract

The average potassium intake in the United States population ranges from 90 to 120 mEq/day. About 98% of the total body's potassium is intracellular, and only 2% is present in the extracellular compartment. This distributional proportion is essential for cellular metabolic reactions and maintaining a gradient for resting membrane potential. A loss of this gradient results in hyper- or hypopolarization of the cell membrane, especially in cardiac muscles leading to life-threatening arrhythmias. Multiple mechanisms in human maintain homeostasis. Transient initial changes are due to transcellular shifts activating sodium-potassium ATPase pumps on the cell membrane. The kidneys essentially take part in excess potassium excretion, maintaining total body stores constant within normal range. Gastrointestinal secretion of potassium is insignificant in individuals with normal renal function, however plays an essential role in individuals with compromised renal function. So far, a classic feedback mechanism was thought to maintain potassium homeostasis; however, a recently recognized feedforward mechanism acting independently also helps preserve potassium homeostasis. Hence, potassium homeostasis is vital for humans to function at a normal level.

Keywords: potassium, homeostasis, gradient, renal, compartment

1. Introduction

The total body potassium (K^+) in an average 70-kg adult is approximately 3000–4000 mEq (50–55 mEq/kg) [1]. About 98% of this is intracellular, approximating to a concentration of 140 mEq/L and 2% in the extracellular compartment, which amounts to 4–5 mEq/L. Potassium's two primary physiologic functions are cellular metabolism, protein, glycogen, deoxyribonucleic acid (DNA) synthesis, and resting potential membrane maintenance. Multiple physiologic processes have been identified in humans that maintain potassium homeostasis with a goal of appropriate tissue potassium distribution [1]. They can be classified into two groups based on the mechanisms involved: transcellular shifts and potassium excess excretion.

Transcellular mechanisms maintain the ICF (intracellular fluid): ECF (extracellular fluid) potassium ratio by acting immediately within the first few minutes to hours by regulating Na^+/K^+-ATPase (sodium/potassium adenosine triphosphatase pump), resulting in transcellular shifts. It is an electrogenic pump transporting sodium and potassium in the ratio of three sodium to two potassium. It is an integral membrane protein and serves as an ion channel, maintaining the electrochemical gradient across the cell membrane. The delayed mechanisms are slower to kick in but play a significant role in the excretion of the excess potassium from the body *via* renal and gastrointestinal mechanisms.

2. Potassium homeostasis mechanisms

2.1 Factors responsible for the transcellular shift of potassium

2.1.1 Insulin

Insulin is released after a meal when the plasma glucose concentration increases. Insulin plays a vital role in shifting dietary potassium into the cells and avoiding hyperkalemia before the kidneys can work on the excretion of the extra potassium [2, 3]. Insulin attaches to specific cell surface receptors and inserts GLUT4 (glucose transporter type-4) into the cell membranes, promoting glucose uptake in insulin-responsive tissues such as skeletal muscles, adipocytes, and cardiomyocytes. More than 80% of glucose is transported into the muscle cells. Also seen is an upregulation of the GLUT4-mediated glucose transport in elevated transport needs, like elevated blood glucose during a carbohydrate-rich meal or during increased metabolic demands by skeletal muscles during exercise [4]. The potassium uptake is also increased during this process by increasing the Na^+/K^+-ATPase activity. Insulin is also noted to have differential glucose and K uptake regulation, as noted in a metabolic syndrome where insulin-mediated glucose is compromised, but the cellular K+ uptake usually occurs [5, 6].

2.1.2 Catecholamines

Catecholamines play a critical role in potassium homeostasis. They alter internal K^+ distribution by acting through alpha and beta receptors. Beta-2 receptors enhance K^+ uptake by cells by activating the Na^+/K^+-ATPase pump [7, 8]. This effect appears to be present at basal catecholamine levels [9, 10]. The beta-adrenergic activity also increases insulin secretion from the pancreas by directly stimulating and enhancing glycolysis, which increases blood glucose levels. A stress response results in epinephrine release that causes an acute drop in plasma K^+ by approximately 0.5– 0.6 mEq/L. Alpha-1,2 adrenoreceptors mediate the initial hyperkalemia by activating hepatic calcium-dependent potassium channels [11, 12]. Beta-3 adrenoreceptor stimulation may also affect the plasma potassium [12].

2.1.3 Exercise

Exercise causes an elevation in plasma K^+ levels. The rise is proportional to the exercise intensity. There is an increase of 0.3–0.4 mEq/L with slow walking, 0.7–1.2 mEq/L with moderate exercise, and a max of 2.0 mEq/L with strenuous exercise leading to exhaustion [13–15]. These changes are transient for few minutes and then normalize. This transient increase is due to the release of potassium from the exercising cells. Skeletal muscle cells have ATP-dependent K^+ channels. A reduction of ATP (adenosine triphosphate) levels causes the opening of more K^+ channels. The release of potassium from the muscles causes a local increase in the plasma K^+ concentration, which causes vasodilation, and hence increases blood supply to the muscles during exercise. This mechanism is attenuated in physical conditioning as there is an increase in both cellular K^+ concentration and Na^+/K^+-ATPase activity.

2.1.4 Acid-base balance changes

The changes in acid-base balance cause alterations in K^+ levels to maintain electroneutrality. Hence, such changes are seen only in the setting of acidosis caused by inorganic acids. Electroneutrality is maintained by the movement of K^+ and Na^+ into the ECF, as chloride (Cl^-) movement into the cells is limited. The wide range of variation is possibly due to other mechanisms involved in regulating potassium homeostasis [16, 17].

Skeletal muscles also have another mechanism to regulate intracellular pH *via* Na^+-H^+ exchanger present on the cell membrane to regulate intracellular pH. The exact mechanism is explained in **Figure 1**.

2.1.5 Plasma tonicity

The changes in plasma tonicity also affect K^+ levels. Effective osmoles such as glucose, mannitol, and sucrose accumulate in the ECF, resulting in an osmotic gradient that causes water movement from ICF to ECF [18–20]. This effect leads to a decrease in cell volume. During this process, the intracellular K^+ concentration increases and causes K^+ efflux through K^+ permeable channels. This process is shown in **Figure 2**.

2.1.6 Plasma K^+ concentration

The plasma K^+ concentration itself influences the movement of K^+ in and out of cells *via* passive mechanisms, hence regulating the homeostasis. For example, there is

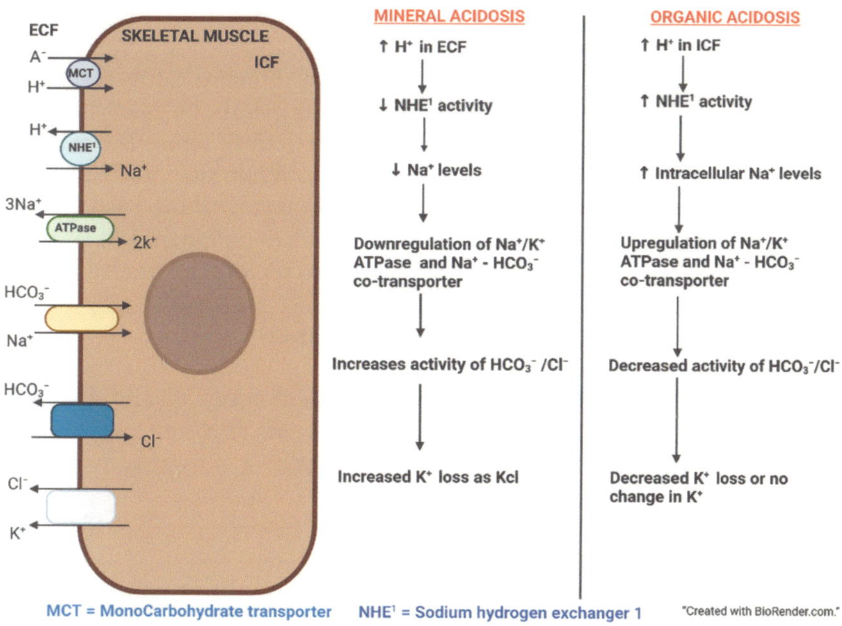

Figure 1.
Intracellular pH regulation by skeletal muscles.

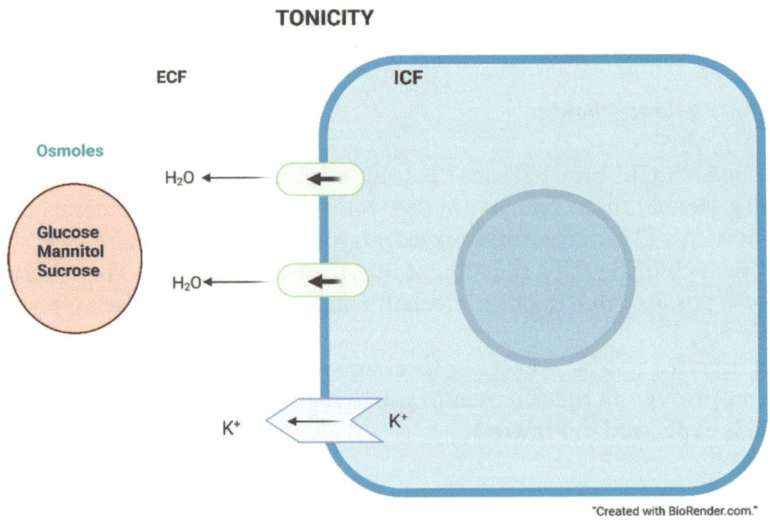

Figure 2.
Plasma tonicity.

an initial elevation of plasma K^+ concentration after a K load, promoting K^+ movement inside the cells. Similarly, in K^+ losses either *via* gastrointestinal or renal processes, there is an initial fall in plasma K^+ level in the plasma within the required normal range transiently, unless the total body K^+ stores are significantly affected [21, 22].

2.1.7 Effect of cell breakdown and production on plasma potassium

Clinical conditions such as severe trauma, tumor lysis syndrome, acute tissue ischemia, and necrosis cause cell breakdown and release of intracellular K^+ into the ECF [23]. The hyperkalemia severity is dependent on the ability of other cells to uptake the excess K^+ and the capability of the kidney to excrete K^+ quickly. Conversely, there is a movement of ECF K^+ to ICF due to increased cellular metabolic needs like protein synthesis in situations with increase in rapid cell production. This effect is observed in cases of severe vitamin B 12 or folic acid deficiency. When such individuals are supplemented with vitamin B 12 or folic acid, there is a marked increase in erythropoiesis and red blood cell production, causing hypokalemia; hence, it is recommended to monitor labs and supplement potassium to replete the levels [24].

2.2 Factors responsible for potassium excretion (renal)

The kidney plays a significant role in the excretion of excess K^+ and maintaining K^+ balance. The colon plays a trivial role as small amounts of K^+ are lost through feces each day. Insignificant amounts are lost through sweat.

2.2.1 Potassium handling in proximal convoluted tubule (PCT)

Around 80% of the filtered potassium is reabsorbed in the proximal collecting tubule. The movement of K^+ in the PCT is mainly passive, following sodium and water. The Na^+ absorption increases across the PCT, by both active and passive processes causing net fluid reabsorption. This process drives K^+ reabsorption by solvent drag. There is also a minimal change in the transepithelial voltage from negative to slightly positive as we move down the PCT, favoring K^+ and calcium (Ca^{2+}) reabsorption *via* the paracellular pathway (**Figure 3**) [2].

2.2.2 Potassium handling in thick ascending limb

K^+ reabsorption in this segment occurs by both paracellular and transcellular pathways. $Na^+/K^+/Cl^-$ co-transporter plays a significant role in the reabsorption of potassium here. This channel is located on the luminal side. The low-intracellular Na^+ concentration and high-transcellular Na^+ gradient caused by the Na^+/K^+-ATPase pump activity in the basolateral membrane help activate the $Na^+/K^+/Cl^-$ co-transporter. This electroneutral process causes an increase in intracellular K^+, which then exits *via* the K^+/Cl^- co-transporter located in the basolateral membrane. A ROMK (renal outer medullary potassium channel) is present on the luminal side, allowing K^+ out of the cell into the lumen. This activity helps maintain the $Na^+/K^+/Cl^-$ co-transporter activity by recycling the K^+ (**Figure 4**) [2, 25].

2.2.3 Potassium handling in distal convoluted tubule (DCT) and collecting duct

Less than 10% of the filtered load reaches the distal tubule. The lumen's K^+ concentration increases down the lumen and is secondary to voltage-dependent K^+ secretion mediated by the ROMK channel. Late DCT has an epithelial Na^+ channel (ENaC) responsible for sodium absorption and creating lumen-negative electrochemical gradient and hence effusion of K^+, similar to collecting duct [26].

Most potassium secretion is in the connecting segment and the collecting tubules (cortical, outer medullary, and inner medullary). This secretion varies according to the physiologic needs of the body. The connecting tubule has two major types of cells: principal cells and intercalated cells.

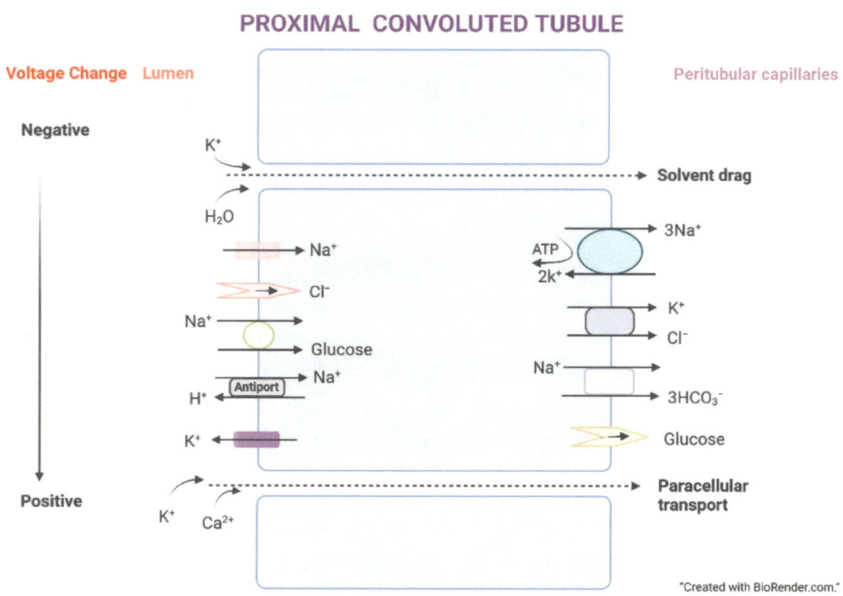

Figure 3.
Potassium handling in proximal convoluted tubule (PCT).

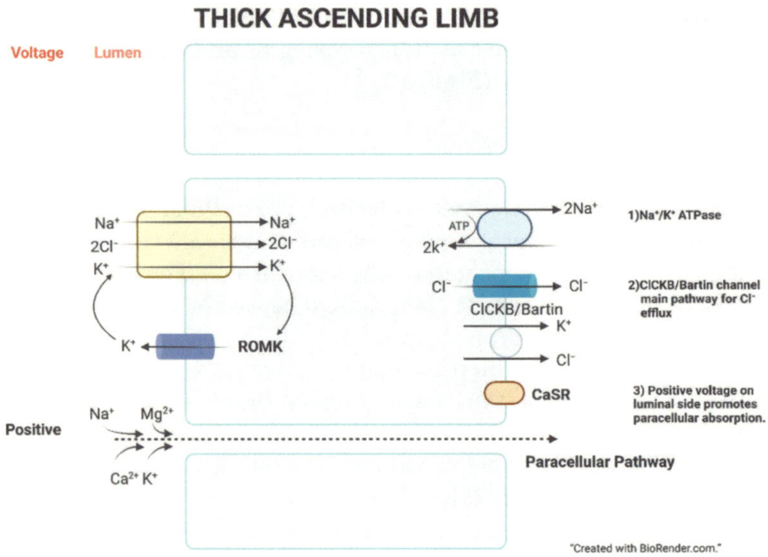

Figure 4.
Potassium handling in the thick ascending limb.

In principal cells, K^+ movement out of the cell into the tubular lumen is due to the electrochemical gradient generated by the entry of Na^+ into the cell *via* the ENaC channel located on the luminal side. The Na^+/K^+-ATPase pump at the basolateral membrane moves 3 Na^+ out of the cells in exchange for 2 K^+ creating a favorable concentration gradient. The gradient leads to the movement of Na^+ from the tubular lumen into the cell. This movement creates a negative voltage on the luminal side. This electrochemical gradient favors K^+ secretion into the lumen. It also promotes paracellular reabsorption of chloride (**Figure 5**) [27].

Intercalated cells are of two types, type A and type B, involved in acid-base regulation. In type A cells, the H^+/K^+ATPase pump on the luminal side results in

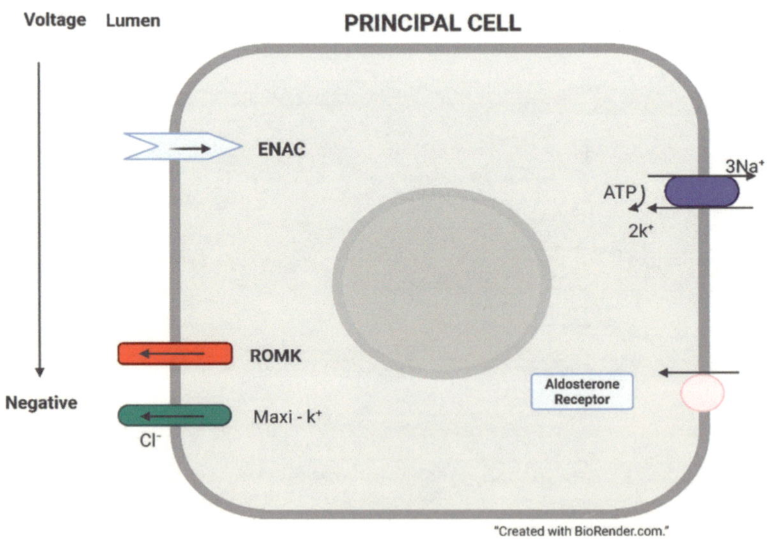

Figure 5.
Potassium handling in principal cells.

H+ secretion and K+ reabsorption (**Figure 6**) [28, 29]. The activity of this pump is increased in K^+-depleted states and decreased in the setting of elevated K^+ levels.

In intercalated type B cells, the reabsorption of K^+ is along with the Cl^- reabsorption *via* Cl^-/HCO_3^- exchanger, which is a proposed mechanism and not yet substantially proven (**Figure 7**) [29–31].

2.3 Role of aldosterone in potassium homeostasis

Aldosterone is a steroid hormone. It enters the cell, binds with a cytosolic receptor and moves to the nucleus, increasing the synthesis of sgk mRNA (serum and glucocorticoid regulated kinase messenger ribonucleic acid), resulting in sgk protein [32].

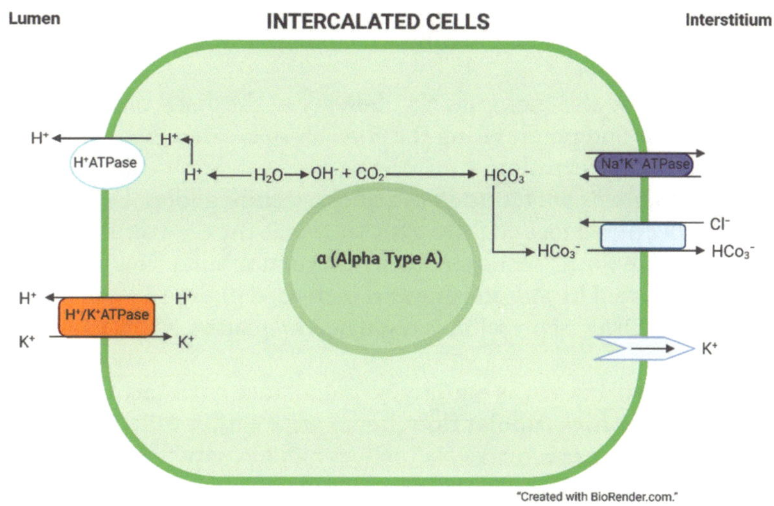

Figure 6.
Potassium handling in intercalated type A cell.

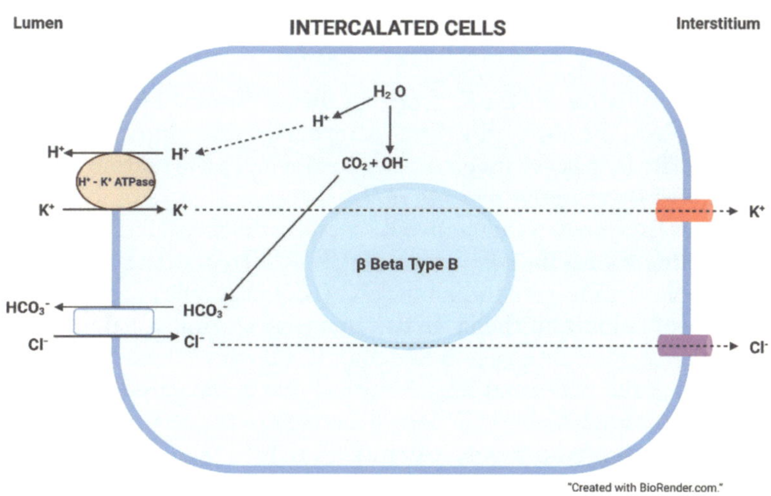

Figure 7.
Potassium handling in intercalated type B cell.

sgk protein stimulates the ENaC activity sevenfold. An increase in serum K^+ levels as little as 0.1–0.2 mEq/L can stimulate significant aldosterone levels [33].

2.4 Tubular flow and potassium secretion

Increasing the distal tubular flow rate potentiates K^+ secretion in those segments. This effect is more prominent when a person consumes a high K^+ diet as it also causes a simultaneous increase in aldosterone levels. Under normal circumstances, the fluid entering the distal tubule has a K^+ concentration of <1 mEq/L as most of the K^+ filtered is absorbed in the earlier parts of the nephron. Due to water reabsorption in the presence of ADH (antidiuretic hormone) and K^+ secretion in the tube's distal parts, the tubular K^+ concentration increases. However, if the distal flow rate is increased, enhanced flow washes the secreted K^+, keeping the K^+ concentration in the tubular fluid relatively lower and creating a favorable concentration gradient for K^+ secretion [34, 35].

Increased distal flow also increases Na^+ delivery to the distal nephron, increasing Na entry into the cells and potentiating the changes to create a favorable electrochemical gradient for K^+ secretion.

Two major K^+ channels are found in the thick ascending loop, DCT, cortical, and medullary collecting ducts. ROMK, also known as the constitutive K^+ secretary channel, is responsible for K^+ secretion during normal tubular flow. Cortical ROMK expression is upregulated by aldosterone and increased plasma K^+ concentration, whereas medullary ROMK expression is regulated by plasma K^+ levels and not aldosterone.

BK channels are also known as big/large conductance channels or "Maxi K" channels opened by the high tubular flow, hence were earlier called a flow-dependent channel. High tubular flow increases Na^+ delivery and eventually causes apical membrane depolarization. The depolarization leads to increased intracellular calcium levels and activates the BK channels. Recent studies suggest that K^+ secretion by ROMK is also increased in the setting of increased tubular flow. Both these channels work together in maintaining potassium homeostasis and preventing hyperkalemia. Either of them is upregulated in the absence of the other channel [36].

2.5 Role of the colon in potassium homeostasis

The primary absorption site for K^+ from the diet is the small intestine. The colon plays a minimal role in the absorption and secretion of potassium. Potassium secretion occurs primarily by passive mechanisms; however, in the rectum and sigmoid colon, K^+ is secreted by an active process [37].

Animal studies have shown that active K^+ secretion is mediated by apical K^+ channels, which coordinate with the basolateral $Na^+/K^+/Cl^-$ co-transporter. Intermediate conductance K^+ (IK) and large-conductance K^+ (BK) channels are present on the apical membrane of colonic epithelia. In patients with chronic renal insufficiency, especially when creatinine clearance is less than 10 mL/min, the net colonic K^+ secretion increases compared to normal renal function due to increased expression and activity of the BK channel. Colonic K^+ losses increase in the setting of diarrhea and by using cation resins (e.g., sodium polystyrene sulfonate) [38].

2.6 Feedforward control of potassium balance

Feedforward control means the response to a specific signal is preset and happens irrespective of the changes in the environment, unlike in the feedback mechanism where the changes in the environment control the pathway.

An increase in dietary intake of K^+ causes an increase in renal excretion of potassium, even though the K^+ concentration is not sufficient to cause any changes in plasma K^+ concentration or stimulate aldosterone. It was also noted that this mechanism acts independently and is not altered by changing the tubular flow rate of urine, urine pH, renal Na^+ excretion, or an aldosterone antagonist. `The feedforward mechanism is wholly dissociated from the common pathways involved in the feedback mechanisms of K^+ homeostasis [39].

The precise mechanism is still unknown, but it is hypothesized that there is a possible gastrointestinal-renal signaling pathway in humans responsible for this feedforward control of K^+ homeostasis [40].

2.7 Circadian rhythm and potassium levels

Potassium excretion is also influenced by the circadian rhythm irrespective of the activity levels or posture. The peak potassium excretion is observed in the middle of the day [41]. The presence of an oscillator system in DCT, connecting tubule and cortical collecting duct renal tubular cells, is responsible for the circadian variation of potassium excretion. The pathway includes cellular receptors for central nervous system signals, intracellular messenger's, effectors, and renal tubule membrane transporters [42]. The oscillations result in circadian gene and transcriptional factor expression changes modifying the expression of vasopressin V2 receptor and multiple transcellular channels (Aquaporin 2, Aquaporin 4, alpha ENaC, and ROMK-1), maintaining the plasma sodium and potassium concentration [39, 43]. ROMK gene expression is higher during physical activity, whereas the H^+/K^+ATPase gene expression is higher at rest [44]. Circadian rhythm causes a variation in aldosterone secretion, thus impacting renal potassium excretion [45].

3. Conclusion

To maintain a normal potassium concentration, both the feedback and feedforward mechanisms work in tandem. After a potassium-rich diet, the increased plasma potassium levels gain entry into the cells *via* transcellular processes. At the same time, the body gears up to activate the renal and the colonic potassium excretion. This is the classical feedback pathway, whereas the feedforward pathway acts independently. Understanding the intricate mechanisms of potassium homeostasis in humans is a must in our clinical approach to potassium disorders for effective treatment strategies.

Acknowledgements

No external funding was received in preparation of this manuscript.

Conflict of interest

We declare no conflict of interest.

Notes/thanks/other declarations

We thank the editor for allowing us to author this manuscript.

Acronyms and abbreviations

K^+	potassium
mEq/L	milliequivalent/liter
kg	kilogram
DNA	deoxyribonucleic acid
ICF	intracellular fluid
ECF	extracellular fluid
Na^+/K^+-ATPase	sodium/potassium-adenosine triphosphatase
Na^+	sodium
GLUT4	glucose transporter type 4
ATP	adenosine triphosphate
Cl^-	chloride
pH	a scale used to specify the acidity or basicity of an aqueous solution
H^+	hydrogen
MCT	monocarbohydrate transporter
NHE1	sodium hydrogen exchanger 1
PCT	proximal convoluted tubule
Ca^{2+}	calcium
$Na^+/K^+/Cl^-$ co-transporter	sodium/potassium/chloride co-transporter
ROMK	renal outer medullary potassium channel
DCT	distal-convoluted tubule
ENaC	epithelial sodium channel
H^+/K^+ATPase	hydrogen/potassium adenosine triphosphatase
Cl^-/HCO_3^- exchanger	chloride/bicarbonate exchanger
mRNA	messenger ribonucleic acid
ADH	antidiuretic hormone
BK/Maxi K channels	big/large conductance channels
IK	intermediate conductance

Author details

Shakuntala S. Patil* and Sachin M. Patil
University of Missouri, Columbia, MO, USA

*Address all correspondence to: drshaku123@outlook.com

References

[1] Palmer BF, Clegg DJ. Physiology and pathophysiology of potassium homeostasis: Core curriculum 2019. American Journal of Kidney Diseases. 2019;**74**(5):682-695

[2] Palmer BF. Regulation of potassium homeostasis. Clinical Journal of the American Society of Nephrology. 2015;**10**(6):1050-1060

[3] Masharani U, German MS. Pancreatic hormones and diabetes mellitus. In: Gardner DG, Shoback D, editors. Greenspan's Basic & Clinical Endocrinology. 10th ed. New York, NY: McGraw-Hill Education; 2017

[4] Vargas E, Podder V, Carrillo Sepulveda MA. Physiology, Glucose Transporter Type 4. Treasure Island (FL): StatPearls; 2021

[5] Nguyen TQ, Maalouf NM, Sakhaee K, Moe OW. Comparison of insulin action on glucose versus potassium uptake in humans. Clinical Journal of the American Society of Nephrology. 2011;**6**(7): 1533-1539

[6] Alvestrand A, Wahren J, Smith D, DeFronzo RA. Insulin-mediated potassium uptake is normal in uremic and healthy subjects. American Journal of Physiology. 1984;**246**(2 Pt 1): E174-E180

[7] Clausen T, Everts ME. Regulation of the Na,K-pump in skeletal muscle. Kidney International. 1989;**35**(1):1-13

[8] Clausen T, Flatman JA. Effects of insulin and epinephrine on Na+-K+ and glucose transport in soleus muscle. American Journal of Physiology. 1987;**252**(4 Pt 1):E492-E499

[9] DeFronzo RA, Bia M, Birkhead G. Epinephrine and potassium homeostasis. Kidney International. 1981;**20**(1):83-91

[10] Brown MJ, Brown DC, Murphy MB. Hypokalemia from beta2-receptor stimulation by circulating epinephrine. New England Journal of Medicine. 1983;**309**(23):1414-1419

[11] Struthers AD, Whitesmith R, Reid JL. Prior thiazide diuretic treatment increases adrenaline-induced hypokalaemia. Lancet. 1983;**1**(8338): 1358-1361

[12] Moratinos J, Reverte M. Effects of catecholamines on plasma potassium: The role of alpha- and beta-adrenoceptors. Fundamental & Clinical Pharmacology. 1993;**7**(3-4):143-153

[13] Sassard J, Vincent M, Annat G, Bizollon CA. A kinetic study of plasma renin and aldosterone during changes of posture in man. The Journal of Clinical Endocrinology & Metabolism. 1976;**42**(1):20-27

[14] Struthers AD, Quigley C, Brown MJ. Rapid changes in plasma potassium during a game of squash. Clinical Science (London). 1988;**74**(4):397-401

[15] Lindinger MI, Heigenhauser GJ, McKelvie RS, Jones NL. Blood ion regulation during repeated maximal exercise and recovery in humans. American Journal of Physiology. 1992;**262**(1 Pt 2):R126-R136

[16] Adrogue HJ, Madias NE. Changes in plasma potassium concentration during acute acid-base disturbances. The American Journal of Medicine. 1981;**71**(3):456-467

[17] Magner PO, Robinson L, Halperin RM, Zettle R, Halperin ML. The plasma potassium concentration in metabolic acidosis: A re-evaluation. American Journal of Kidney Diseases. 1988;**11**(3):220-224

[18] Kunin AS, Surawicz B, Sims EA. Decrease in serum potassium concentrations and appearance of cardiac arrhythmias during infusion of potassium with glucose in potassium-depleted patients. New England Journal of Medicine. 1962;**266**:228-233

[19] Nicolis GL, Kahn T, Sanchez A, Gabrilove JL. Glucose-induced hyperkalemia in diabetic subjects. Archives of Internal Medicine. 1981;**141**(1):49-53

[20] Moreno M, Murphy C, Goldsmith C. Increase in serum potassium resulting from the administration of hypertonic mannitol and other solutions. The Journal of Laboratory and Clinical Medicine. 1969;**73**(2):291-298

[21] DeFronzo RA, Lee R, Jones A, Bia M. Effect of insulinopenia and adrenal hormone deficiency on acute potassium tolerance. Kidney International. 1980;**17**(5):586-594

[22] Sterns RH, Cox M, Feig PU, Singer I. Internal potassium balance and the control of the plasma potassium concentration. Medicine (Baltimore). 1981;**60**(5):339-354

[23] Drakos P, Bar-Ziv J, Catane R. Tumor lysis syndrome in nonhematologic malignancies. Report of a case and review of the literature. American Journal of Clinical Oncology. 1994;**17**(6):502-505

[24] Lawson DH, Murray RM, Parker JL. Early mortality in the megaloblastic anaemias. QJM. 1972;**41**(161):1-14

[25] Burg MB. Thick ascending limb of Henle's loop. Kidney International. 1982;**22**(5):454-464

[26] Subramanya AR, Ellison DH. Distal convoluted tubule. Clinical Journal of the American Society of Nephrology. 2014;**9**(12):2147-2163

[27] Wang WH, Schwab A, Giebisch G. Regulation of small-conductance K+ channel in apical membrane of rat cortical collecting tubule. American Journal of Physiology. 1990;**259**(3 Pt 2): F494-F502

[28] Garg LC. Respective roles of H-ATPase and H-K-ATPase in ion transport in the kidney. Journal of the American Society of Nephrology. 1991;**2**(5):949-960

[29] Weiner ID, Wingo CS. Hyperkalemia: A potential silent killer. Journal of the American Society of Nephrology. 1998;**9**(8):1535-1543

[30] Zhou X, Xia SL, Wingo CS. Chloride transport by the rabbit cortical collecting duct: Dependence on H, K-ATPase. Journal of the American Society of Nephrology. 1998;**9**(12):2194-2202

[31] Madsen KM, Verlander JW, Tisher CC. Relationship between structure and function in distal tubule and collecting duct. Journal of Electron Microscopy Technique. 1988;**9**(2):187-208

[32] Chen SY, Bhargava A, Mastroberardino L, Meijer OC, Wang J, Buse P, et al. Epithelial sodium channel regulated by aldosterone-induced protein sgk. Proceedings of the National Academy of Sciences of the United States of America. 1999;**96**(5):2514-2519

[33] Scott JH, Menouar MA, Dunn RJ. Physiology, Aldosterone. Treasure Island (FL): StatPearls; 2021

[34] Rodan AR, Huang CL. Distal potassium handling based on flow modulation of maxi-K channel activity. Current Opinion in Nephrology and Hypertension. 2009;**18**(4):350-355

[35] Wald H, Garty H, Palmer LG, Popovtzer MM. Differential regulation of ROMK expression in kidney cortex and medulla by aldosterone and potassium. American Journal of Physiology. 1998;**275**(2):F239-F245

[36] Sansom SC, Welling PA. Two channels for one job. Kidney International. 2007;**72**(5):529-530

[37] Agarwal R, Afzalpurkar R, Fordtran JS. Pathophysiology of potassium absorption and secretion by the human intestine. Gastroenterology. 1994;**107**(2):548-571

[38] Sandle GI, Gaiger E, Tapster S, Goodship TH. Enhanced rectal potassium secretion in chronic renal insufficiency: Evidence for large intestinal potassium adaptation in man. Clinical Science. 1986;**71**(4):393-401

[39] Epstein M, Lifschitz MD. Potassium homeostasis and dyskalemias: The respective roles of renal, extrarenal, and gut sensors in potassium handling. Kidney International Supplements 2016;6(1):7-15.

[40] McDonough AA, Youn JH. Potassium homeostasis: The knowns, the unknowns, and the health benefits. Physiology (Bethesda). 2017;**32**(2):100-111

[41] Moore Ede MC, Brennan MF, Ball MR. Circadian variation of intercompartmental potassium fluxes in man. Journal of Applied Physiology. 1975;**38**(1):163-170

[42] Gumz ML, Rabinowitz L. Role of circadian rhythms in potassium homeostasis. Seminars in Nephrology. 2013;**33**(3):229-236

[43] Zuber AM, Centeno G, Pradervand S, Nikolaeva S, Maquelin L, Cardinaux L, et al. Molecular clock is involved in predictive circadian adjustment of renal function. Proceedings of the National Academy of Sciences of the United States of America. 2009;**106**(38):16523-16528

[44] Salhi A, Centeno G, Firsov D, Crambert G. Circadian expression of H,K-ATPase type 2 contributes to the stability of plasma K(+) levels. The FASEB Journal. 2012;**26**(7):2859-2867

[45] Doi M, Takahashi Y, Komatsu R, Yamazaki F, Yamada H, Haraguchi S, et al. Salt-sensitive hypertension in circadian clock-deficient Cry-null mice involves dysregulated adrenal Hsd3b6. Nature Medicine. 2010;**16**(1):67-74

Chapter 3

Potassium Derangements: A Pathophysiological Review, Diagnostic Approach, and Clinical Management

Sairah Sharif and Jie Tang

Abstract

Potassium is an essential cation critical in fluid and electrolyte balance, acid–base regulation, and neuromuscular functions. The normal serum potassium is kept within a narrow range of 3.5–5.2 meq/L while the intracellular concentration is approximately 140–150 meq/L. The total body potassium is about 45–55 mmol/kg; thus, a 70 kg male has an estimated ~136 g and 60 kg female has ~117 g of potassium. In total, 98% of the total body potassium is intracellular. Skeletal muscle contains ~80% of body potassium stores. The ratio of intracellular to extracellular potassium concentration (Ki/Ke) maintained by Na^+/K^+ ATPase determines the resting membrane potential. Disturbances of potassium homeostasis lead to hypo- and hyperkalemia, which if severe, can be life-threatening. Prompt diagnosis and management of these problems are important.

Keywords: potassium, transcellular shift, renal excretion, hypokalemia, hyperkalemia

1. Introduction

1.1 Cellular shift of potassium

The body maintains potassium (K) homeostasis by two key mechanisms: transcellular shift and renal K reabsorption/excretion [1–7]. The transcellular shift acts immediately within minutes to hours in response to K disturbances in the extracellular fluid. This is also called internal K balance. Cellular shifting is extremely important in the body's defense against K disturbances in the extracellular space. Without transcellular redistribution, even small disturbances in K balance could lead to life-threatening potassium derrangements.

Important internal K regulators are catecholamines, insulin, thyroid hormone, tonicity, mineral acidosis, and various medications [8–10]. Insulin binds to cellular receptors, leading to increased activity of glucose transporter (GLUT4) and Na^+/K^+ ATPase, while catecholamines (via beta2 adrenergic receptors) also upregulate

IntechOpen

Na^+/ K^+ ATPase. This in turn causes uptake of K into cells. Alpha-adrenergic receptor stimulation shifts K into the extracellular space; however, under physiological conditions, this effect is less significant. Theophylline and caffeine exert the same effect by increasing the activity of Na^+/ K^+ ATPase via inhibition of cellular phosphodiesterase and degradation of cyclic adenosine monophosphate (cAMP). There are some important blockers of cellular K channels that prohibit exit of K from cells, causing severe life-threatening hypokalemia. Examples of these entities include: chloroquine, verapamil, barium, and cesium [11–16]. Rapidly dividing cells with high metabolic activity, as seen during initiation of therapy for megaloblastic anemia, could cause large K shifts into cells and consequently hypokalemia [16]. Extracellular mineral acidosis downregulates Na^+/H^+ exchanger 7 [7, 9]. This downregulates the activity of Na^+/K^+ ATPase and ultimately K released out of cells. Metabolic alkalosis has the opposite effect with K shifting intracellularly. Finally, changes in plasma osmolality also lead to cellular K shifts.

1.2 Periodic paralysis

Periodic paralysis (PP) includes a group of rare neuromuscular disorders characterized by episodic paralysis. It includes three main phenotypes: hypokalemic periodic paralysis, hyperkalemic periodic paralysis, and thyrotoxic periodic paralysis. Both genetic and acquired forms of the disorder have been described. The genetic form is usually caused by mutations in the calcium channel (CACNA1S, 60%), sodium channel (SCN4A, 20%), and less commonly inward rectifying potassium channels (KCNJ2, KCNJ18) [17–20]. The genetic mutations alone would not cause K to shift, but instead sensitize skeletal muscle to changes in serum K [21]. In hypokalemic PP symptoms are triggered after there is a drop in K due to cellular shifting such as a carbohydrate rich meal, exercise, or K losses. In hyperkalemic PP symptoms are usually triggered after K-rich food or severe exercise. Acquired thyrotoxic periodic paralysis is more prevalent in Asian populations from China, Taiwan, Japan, and Philippines. Patients are generally young males with a history of recurrent partial or complete paralysis, with a tendency to affect lower limbs more than upper limbs. The episodes may also be precipitated by exercise or carbohydrate-rich meals. The condition is triggered by transcellular K shift in response to an enhanced catecholamine action mediated by thyroxine [22].

1.3 Renal handling of potassium

Potassium is freely filtered across the glomerular membrane (**Figure 1**). Approximately 90% of K is reabsorbed by the early part of the tubule and less than 10% reaches the distal nephron. The bulk of the reabsorption takes place in the proximal convoluted tubule (PCT), which accounts for ~65–70% of the filtered K, followed by the thick ascending limb (TAL), which accounts for ~25%. The more distal nephron reabsorbs or secretes K based on the body requirements, thus playing a critical role in maintaining K balance [23, 24].

In the PCT, the Na^+/ K^+ ATPase drives the active sodium reabsorption. This causes net inward fluid movement, including passive reabsorption of K with water due to solvent drag. In the ascending limb of the loop of Henle, the main channels are apical sodium/potassium/chloride co-transporter (NKCC2), apical renal outer medullary potassium channel (ROMK), and basolateral Na^+/ K^+ ATPase. The reabsorption of K takes place via NKCC2 [25]. The distal nephron consists of early distal convoluted

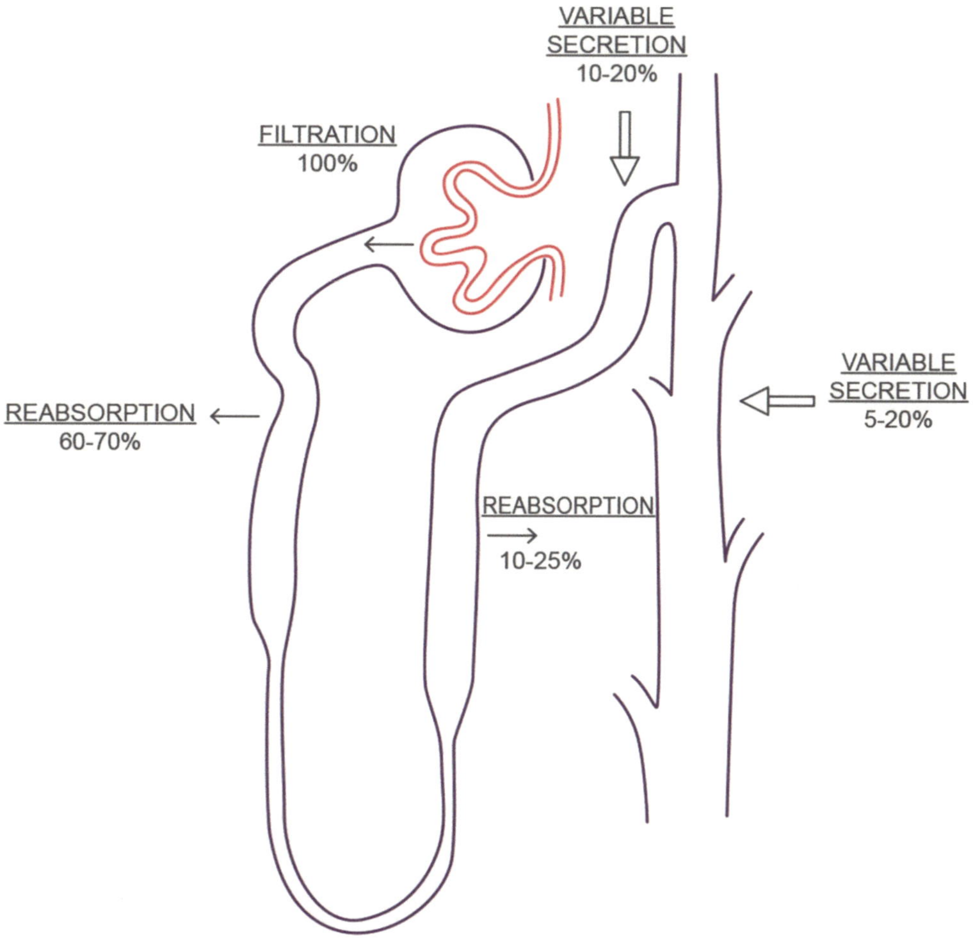

Figure 1.
Schematic representation of renal tubular handling of potassium. Urinary potassium is excreted after filtration, reabsorption, and secretion along the tubules. Over 90% of reabsorption of potassium takes place in the PCT and TAL, whereas DCT, cortical and medullary CD have capacity for variable reabsorption and secretion. PCT proximal convoluted tubule, TAL thick ascending limb, DCT distal convoluted tubule, CD collecting duct.

tubule (DCT1), late distal convoluted tubule (DCT2), connecting tubule (CNT), and collecting duct (CD). DCT1 contains apical Na^+ Cl^- cotransporter channel (NCC), basolateral Na^+/ K^+ ATPase, and basolateral heteromeric K^+ channel (Kir 4.1/ 5.1), among others. DCT2 is a transition between DCT1 and CNT with epithelial sodium channel (ENaC) expressions. NCC is regulated by two proteins from the serine/threonine kinase family called with-no-lysine (WNK) 1 and 4 [26], which in turn are regulated by Kir 4.1/ 5.1. Loss of function mutation in Kir 4.1/ 5.1 depolarizes DCT, inactivates NCC, and results in salt wasting. This will modify ENaC action downstream and affect K secretion [27]. Recent studies indicate that Kir 4.1/5.1 can sense dietary K changes. A high K diet can inhibit Kir 4.1/ 5.1, whereas a low K diet activates it [28]. As such, Kir 4.1/ 5.1 plays an important role in K homeostasis via sensing K intake and modulating NCC and ENaC activities [29, 30]. In the collecting duct (CD), there are two types of cells—principal cells and intercalated cells (alpha and beta). The main channels involved are basolateral Na^+/K^+ ATPase, apical ENaC, ROMK, and Maxi-K (flow dependent) channels. Aldosterone binds to its mineralocorticoid receptor (MR) and stimulates ENaC expression [31]. The net result is enhanced Na reabsorption that causes luminal electronegativity and drives K excretion via ROMK channel. The increased urine flow in this nephron segment can also enhance K loss via Maxi-K.

1.4 Aldosterone paradox

Aldosterone paradox [32] refers to the ability of aldosterone to stimulate reabsorption of sodium without excessive secretion of K in the setting of volume depletion and its ability to stimulate K excretion without sodium retention during hyperkalemia and euvolemia. Volume depletion leads to an activation of the renin angiotensin aldosterone system (RAAS) and a drop in glomerular filtration rate (GFR). The reduced urinary flow and Na delivery will limit the amount of K secretion. Furthermore, angiotensin II (AT II) can directly inhibit ROMK activity when there is a K deficit [33]. In hyperkalemia, distal delivery of sodium and urine volume is preserved due to the lack of AT II stimulation, allowing sufficient K secretion stimulated by the increase in aldosterone [34]. As a result, excessive fluid retention is not present.

1.5 Adaptation to hyperkalemia

Due to the renal adaptive responses, normal extracellular K levels are usually maintained even when there is a significant fall in GFR [35]. Hyperkalemia only develops when there is a severe defect in the distal nephron. This is due to "Potassium Adaptation" from changes in the remaining intact distal nephron, where it undergoes an adaptive increase in expressions of ROMK, ENaC, and Na^+/K^+ ATPases in order to maintain body K homeostasis. This change is in part due to higher aldosterone levels and will achieve a new steady state. However, it should be noted this steady state in chronic kidney disease (CKD) is delicate and can be easily disrupted again with an increased K intake or use of RAAS inhibitors [36].

Under normal circumstances, the gastrointestinal tract, responsible for 5–10% of K excretion, plays a minimal role in K balance. However, it can adapt in times of K derangement by increasing K excretion in cases of advancing CKD. Studies indicate at cellular level, this is due to increased expression of Na^+/K^+ATPase and the apical colonic BK channels 41 [37, 38]. This gastrointestinal adaptation plays an instrumental role in maintaining K homeostasis in end stage renal disease (ESRD) patients [39, 40].

1.6 Renal circadian rhythm

The kidneys play a major role in regulating K balance, excreting ~90% of the dietary K load [41, 42]. The renal K excretion follows a circadian rhythm regulated by central (suprachiasmatic nucleus) and peripheral (renal cells) biological clocks [43]. Full details of this control mechanism are still poorly defined and are beyond the scope of this chapter. Suffice to say that even with destruction of suprachiasmatic nucleus, the kidneys are able to maintain a circadian excretion of K. Elegant studies have demonstrated that maximum kaliuresis occurs at noon, reaches a nadir at midnight, and rhythm is maintained with a high K diet as well [44].

1.7 Pseudo-hyper and hypokalemia

Pseudohyperkalemia is defined as an elevation of serum K by more than 0.3 meq/L over the plasma K [45]. Pseudohyperkalemia can result from either mechanical trauma of cells during phlebotomy [46] or a transcellular shift of K out of cells in the test tube. It is commonly seen when there is extremely elevated leukocyte [47], erythrocyte, or thrombocyte counts [48]. One study suggested that every 100,000/ml rise in platelet count correlated with a K rise by 0.15 meq/L. There is a rare autosomal dominant disease called familial pseudohyperkalemia where at lower temperatures (usually <20°C), red blood cell membranes can leak K leading to pseudohyperkalemia [49]. This can also occur in hereditary spherocytosis [50]. Rarely reverse pseudohyperkalemia can be seen when plasma K is falsely elevated over the serum K. It has been reported in patients with severe leukocytosis and heparin-induced cell lysis [51]. In such cases, arterial blood gas analysis is more accurate.

Pseudohypokalemia can occur in vitro if there is an extremely high leukocyte count (>100,000/ml) or can be temperature-induced [52]. Delayed transport of samples in hot temperatures has been implicated in pseudohypokalemia due to the temperature-mediated stimulation of Na^+/K^+ ATPase [53, 54]. It can be prevented with cold storage of samples at 4 deg. C or if plasma or serum is rapidly separated from cells.

2. Hypokalemia

2.1 Epidemiology

Hypokalemia is defined as serum K level less than 3.5 mEq/L, whereas severe hypokalemia is defined as K level below 2.5 meq/L [55]. Hypokalemia is common with its prevalence reaching 14% in the community setting and 20% among hospitalized patients [56, 57]. In a study of patients with CKD, hypokalemia was associated strongly with an increased mortality (Hazard ratio 1.49, 95% confidence interval 1.26–1.76) [58]. Similar mortality association was observed in another study in patients with CKD and other comorbidity [59].

2.2 Etiopathogenesis

Etiologies of hypokalemia include intracellular shifting, reduced intake, increased excretion, or a combination of these factors. Reduced intake should always be suspected in patients whom are ill with evidence of malnutrition, having eating disorders, or abusing alcohol. Excessive K loss may occur from the gastrointestinal tract, as seen in diarrhea, malabsorption, colonic diseases such as inflammatory bowel disease, and hypersecretory adenomas.

Renal losses could be due to endocrine disorders or tubular cell defects including channelopathies or receptor abnormalities. Proximal tubular defects can lead to a variety of electrolyte problems including hypokalemia and metabolic acidosis. These defects could be inherited (Fanconi syndrome) or acquired in the setting of systemic disease or drug use. In the thick ascending limb, inherited (Bartter syndrome) or acquired defects (i.e., hypercalcemia, loop diuretic use) in NKCC2 or in any other relevant channels/sensors at this nephron segment can also lead to hypokalemia, along with volume depletion and metabolic alkalosis 60 [60–63]. Several transporters orchestrate K reabsorption in the thick ascending limb. Na^+/K^+ ATPase provides the driving force for NKCC2, which reabsorbs sodium, K, and chloride. Additionally, there is a calcium-sensing receptor in the basal-lateral surface, which can inhibit NKCC2 upon stimulation. Similarly, in DCT1, several channels work alongside to promote salt absorption. A defect in NCC will lead to Gitelman syndrome, and its suppression by medications such as thiazide diuretics will share a similar phenotype [64]. The salt wasting present in both Bartter and Gitelman syndrome is associated with an increase in distal urine flow and secondary RAAS stimulation and subsequent K loss.

In the CD, hypokalemia is usually caused by an overactive RAAS axis such as renin tumor, renovascular hypertension, aldosterone oversecretions, and glucocorticoid remediable hypertension (GRA) [65]. De novo activation of ENaC, can also result from non-aldosterone-mediated activation of MR or gain-of-function mutation of MR [66–69]. Finally, drugs such as licorice, carbenoxolone, and gossypol can also lead to an enhanced mineralocorticoid action [16]. Magnesium deficiency in the presence of aldosterone stimulation will exacerbate K losses in the distal nephron via ROMK [70].

2.3 Clinical features

Clinical symptoms depend on the timing and severity of hypokalemia, as well as the presence of certain comorbidities. Symptoms are more evident if serum K falls below 3 meq/L, hypokalemia is relatively rapid, concomitant use of digoxin [71, 72], and ischemic heart disease [16]. Since K is critical for maintaining and modulating resting membrane potential, both cardiac and skeletal muscles can exhibit electrophysiologic changes in response to hypokalemia leading to arrhythmias, paresis, and paralysis [73]. The electrocardiogram (ECG) may show tall P waves, prominent J waves, ST-segment depression, prolonged QT interval, T wave flattening, U wave, premature ventricular contraction, and ventricular tachycardia [74].

Other symptoms of hypokalemia include generalized ascending muscle weakness, pain, and ileus. Severe hypokalemia can also trigger rhabdomyolysis. Renal effects of hypokalemia include nephrogenic diabetes insipidus, ammoniagenesis with subsequent activation of alternative complement pathways [75] resulting in inflammation and fibrosis [76].

3. Diagnostic approach

3.1 Initial evaluation

The clinical investigation of hypokalemia starts with a thorough history and physical examination, followed by laboratory testing including a full metabolic panel (**Figures 2** and **3**). It is important to rule out pseudohypokalemia and assess cellular shifting.

3.2 Renal potassium excretion

The normal kidney responds to hypokalemia by lowering the K excretion. However, even if a patient has no oral intake, the obligatory K excretion is ~15 meq/day [77]. It's useful to quantify renal K excretion (spot or 24 hours) to determine if there is appropriate renal conservation [55].

3.3 Blood pressure and acid: Base status

If a patient has metabolic acidosis, consider differentials of gastrointestinal losses, proximal renal tubular acidosis (pRTA), distal renal tubular acidosis (RTA type 1),

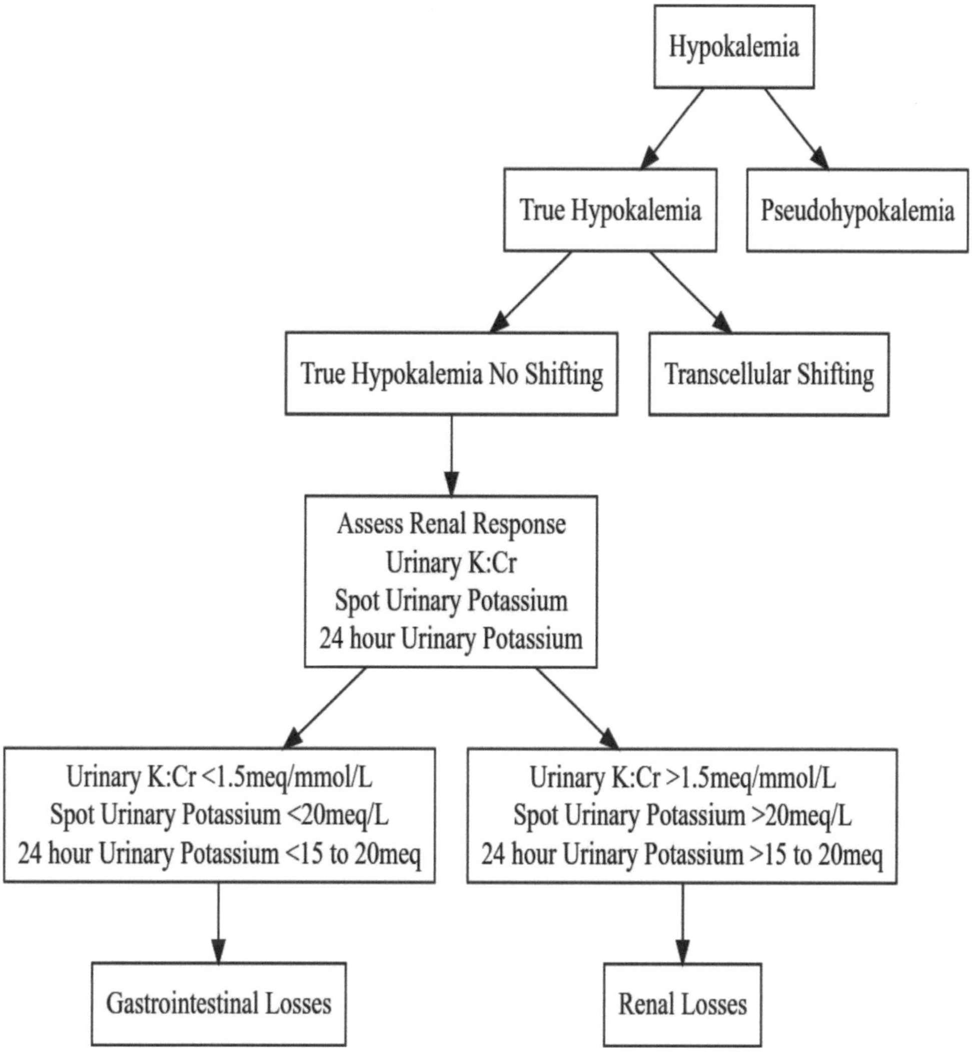

Figure 2.
A diagnostic approach to hypokalemia (part 1).

or diabetic ketoacidosis (DKA) [77]. Of note, typically gastric losses tend to cause metabolic alkalosis while intestinal losses will cause nongap metabolic acidosis. When metabolic alkalosis is present with normal or low blood pressure, the differential diagnosis should include salt wasting syndromes beyond the proximal tubule such as Bartter syndrome, Gitelman syndrome, and diuretic use [55]. When hypertension and metabolic alkalosis are present, plasma renin, aldosterone, and aldosterone-to-renin ratio (ARR) should be checked for a deranged RAAS axis (**Figures 2** and **3**). High renin levels may indicate renin-secreting tumor or renovascular hypertension. When aldosterone is high and renin is suppressed, there is suspicion for hyperaldosteronism or glucocorticoid remediable hypertension (GRA). When both renin and aldosterone are suppressed, there is pathological activation of ENaC via non-aldosterone mechanism (some types of CAH, apparent mineralocorticoid excess (AME), Liddle's syndrome, Geller syndrome, hypercortisolism). Further testing includes checking

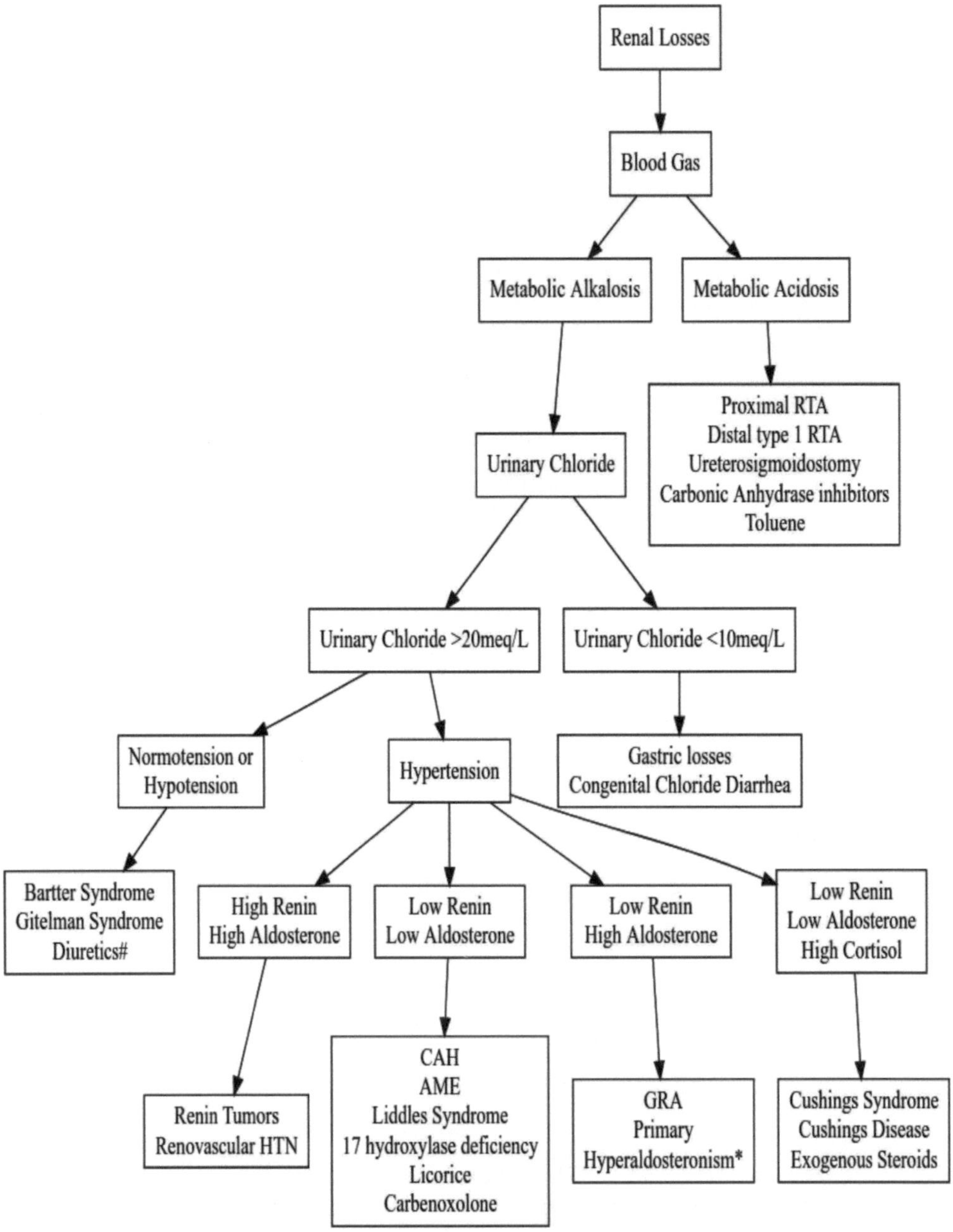

Figure 3.
*A diagnostic approach to hypokalemia (part 2). Abbreviations: RTA renal tubular acidosis, GRA glucocorticoid remediable Aldosteronism, AME apparent mineralocorticoid excess (11 beta hydroxysteroid dehydrogenase deficiency), CAH congenital adrenal hyperplasia (17 alpha hydroxylase, 11 beta hydroxylase deficiency), HTN hypertension. * adrenal adenoma, adrenal hyperplasia, adrenal carcinoma. # loop diuretics, thiazide diuretics.*

the adrenal axis with imaging and endocrine labs (11 beta hydroxylase, or 17 alpha hydroxylase deficiency) [78, 79], cortisol, genetic testing for AME, Liddle syndrome, GRA, and Geller's syndrome [80].

3.4 Management

An ECG is recommended to rule out cardiac dysrhythmias. It is important to treat the underlying cause, reduce losses, and replenish body K stores. Severe hypokalemia warrants closer cardiac monitoring. There is a rough estimation that a 1 meq/L fall in serum K represents a total body K deficit of 200–400 meq [6]. This could be repleted orally or intravenously. Intravenous K repletion warrants cardiac monitoring. It's preferable to give the solution in saline as dextrose causes insulin release and may further drop the serum K. In patients with CKD or ESRD, K repletion is based on cautious evaluation by the nephrology team. Hypomagnesemia should also be corrected if present.

4. Hyperkalemia

4.1 Epidemiology

Hyperkalemia is defined as serum K > 5.2 meq/L. It is also a common and clinically relevant problem. Studies reported 2.5–4% prevalence rates in emergency visits and inpatient hospitalizations in the United States and Canada [81, 82]. As expected, the prevalence rate is much higher in patients with compromised kidney function. In a study of 238 patients with estimated GFR of 15 ml/min/1.73m^2, 31.5% had serum K > 5.5 meq/L, [83]. In another study of men with GFR < 37 ml/min/1.73 m^2, 7.7% had serum K > 5.3 meq/L [84]. Its prevalence is also high (5–40%) in renal transplant recipients due to calcineurin inhibitor use [85]. Like hypokalemia, hyperkalemia is also associated with a higher mortality. In a meta-analysis of over 1.2 million patients with CKD (average GFR 83 ml/min/1.73m^2), serum K concentrations >5.5 meq/L were associated with a hazard ratio of 1.22 for mortality (95% confidence interval 1.15–1.29) [58].

4.2 Etiopathogenesis

Hyperkalemia is caused by excessive intake, extracellular shift or release, and reduced renal excretion. In diabetic ketoacidosis, the relative insulin deficiency along with osmotic forces and extracellular acidosis can lead to hyperkalemia despite total body K depletion [9]. Normal saline can induce hyperchloremic acidosis and cause K shifting [86]. Both rhabdomyolysis and tumor lysis can also result in extensive K release into the extracellular space leading to severe hyperkalemia.

GFR is a critical factor in the renal excretion of K [87]. A GFR < 15 ml/min/1.73 m2 is an important risk factor for hyperkalemia [88]. Patients with diabetes mellitus and RAAS inhibition also have reduced renal K excretion and can develop hyperkalemia. The elimination of K for the most part is controlled via aldosterone, but an elusive aldosterone-independent mechanism also appears to exist [89]. Aldosterone is synthesized by aldosterone synthase (AS) in the adrenal cortex in response to volume depletion and hyperkalemia. Aldosterone deficiency or lack of its action can lead to hyperkalemia. Common causes are hypoaldosteronism (seen in diabetes, Addison's disease, obstructive uropathy, renal tubular acidosis type 4), and pseudohypoaldosteronism (PHA). PHA type I is caused by mutations in MR, whereas PHA type II is caused by mutations in WNK 1 or WNK 4 leading to increased

Mechanism of Action	Examples of Drugs
Blocks ENaC	Amiloride, trimethoprim, pentamidine
Inhibits Renin	DRI, heparin
Inhibits ACE	ACEI
Blocks AR 1	ARB
Inhibit PG Synthesis	NSAID
Na^+/K^+ ATPase	Digoxin
Potassium Leakage from Cells	Succinylcholine
Inhibits Aldosterone Synthesis	CNI, heparin

Table 1.
Drugs causing Hyperkalemia [92, 93].

activation of NCC, reduced distal sodium delivery, and ultimately reduced K secretion via ROMK. Other genetic defects in enzymes involved in cortisol or aldosterone synthesis can also lead to hyperkalemia, i.e., 21 hydroxylase, or aldosterone synthase deficiency [90, 91].

Drugs are important causes of hyperkalemia (**Table 1**). Most of them act on the RAAS axis. Calcineurin inhibitors (CNI) are commonly used in the transplant population and can reduce the expressions of both aldosterone and its receptor [94, 95]. CNIs may also increase NCC activity leading to sodium retention and a reduction in K excretion in the distal nephron [96]. Finally, constipation can also contribute to hyperkalemia especially in patients with end-stage kidney disease. As in those patients, a compensatory increase in colonic K excretion contributes to daily K homeostasis.

4.3 Clinical features

The most important manifestation of hyperkalemia is cardiac dysrhythmia. ECG changes often follow the severity of hyperkalemia. Initially "peaked T wave" is seen due to shortening of depolarization. With progressive worsening of hyperkalemia, PR prolongation, disappearance of P wave, and marked widening of QRS complex will follow [97]. Ultimately patients can develop intraventricular blocks, bradycardia, ventricular arrhythmias [97–100] such as asystole, ventricular fibrillation, pulseless idioventricular rhythm [97] and cardiac arrest [101]. On rare occasions, it can cause myopathy or paralysis [102].

5. Approach to diagnosis

5.1 Initial evaluation

Assessment of hyperkalemia starts with patient history, physical examination, followed by full metabolic panel, blood gas, urine studies, and ECG. Pseudohyperkalemia (**Figure 4**) should be suspected in a patient with abnormally high blood cells. Cellular shifting is mainly due to mineral acidosis, beta blockade, insulin resistance or deficiency. To check for cellular breakdown, serum measurements for creatine phosphokinase and lactate dehydrogenase will be helpful. Rarely patients with normal renal function can develop hyperkalemia if intake is excessive and/or concomitant inhibition of RAAS [103, 104].

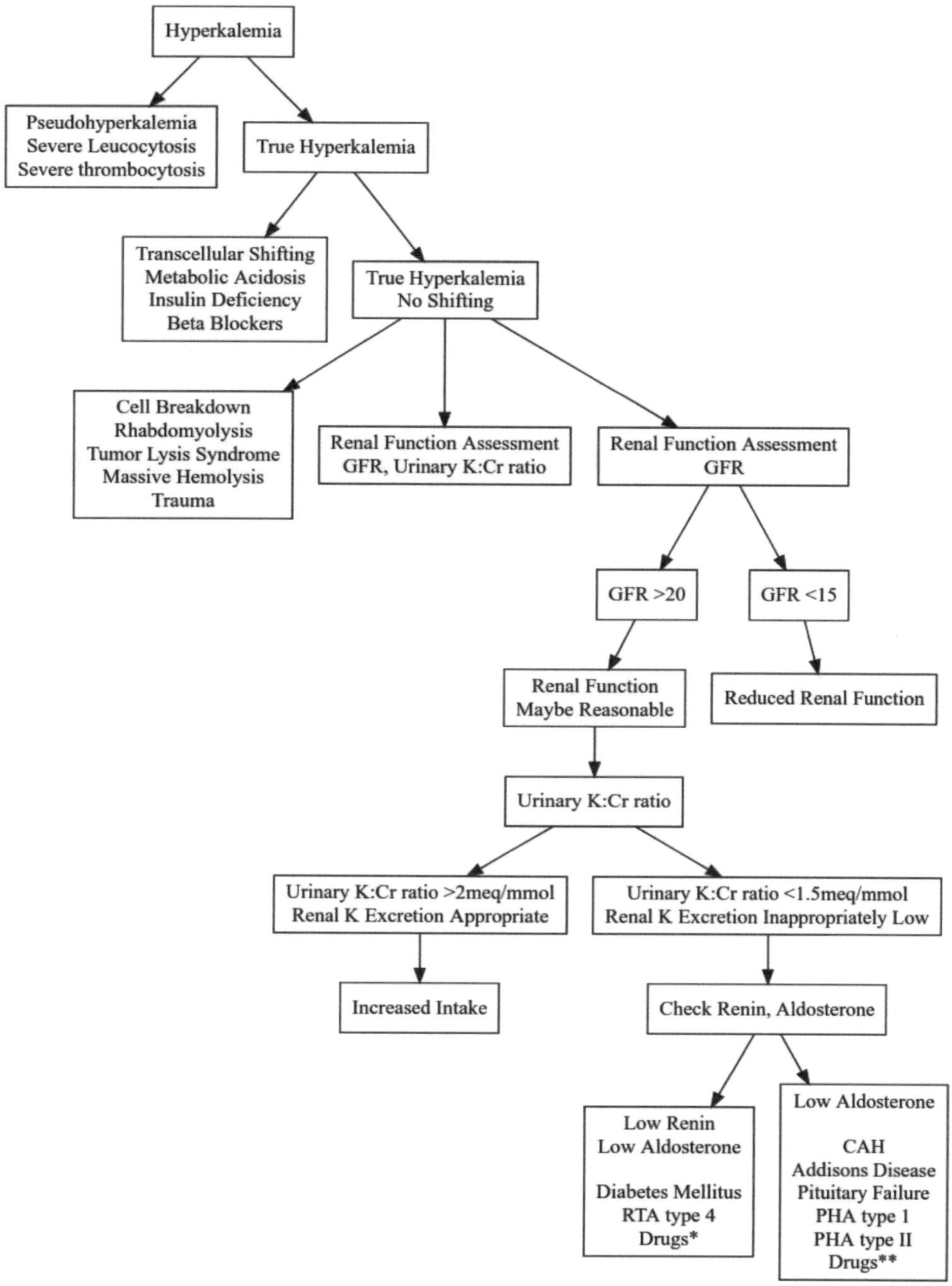

Figure 4.
*A diagnostic approach to hyperkalemia. Abbreviations: RTA renal tubular acidosis, GFR glomerular filtration rate, PHA Pseudohypoaldosteronism, CAH congenital adrenal hyperplasia (21 hydroxylase deficiency). *Beta blocker, alpha 2 agonist, non-steroidal anti-inflammatory agent. **angiotensin converting enzyme inhibitor, angiotensin receptor blocker, renin inhibitor.*

5.2 Renal function and RAAS assessment

To assess the renal K excretion, random urine K, K/Cr ratio, fractional excretion of K, or 24 hr. urine K should be measured. The RAAS axis should also be evaluated by checking the plasma levels of renin and aldosterone. A high aldosterone level indicates a downstream antagonism of aldosterone action, commonly seen with obstructive uropathy, PHA, distal RTA type 4, and the use of certain drugs. If plasma aldosterone level is low, then inherited or acquired causes of hypoaldosteronism should be suspected (**Figure 4**). Further testing can be done to evaluate for adrenal axis, such as ruling out cortisol deficiency with serum cortisol, ACTH, and 21 hydroxylase [88].

5.3 Management

Management of hyperkalemia depends on the severity, underlying cause and signs of serious complication, i.e., high-risk ECG changes. If hyperkalemia is severe and there are significant ECG changes, patients should get intensive care unit monitoring. Hyperkalemia enhances depolarization of cardiac membrane, activates inward rectifying K channels [73, 99]. If a patient has ECG changes, intravenous calcium gluconate should be given to stabilize cardiac membrane potential. Calcium raises cell depolarization threshold, which reduces myocardial excitation [105]. Intracellular shifting of K by insulin, beta agonists, bicarbonate is also helpful.

In addition to restricting intake, medications including RAAS blockers should be temporarily discontinued. Due to their established benefit in renal and cardiovascular disease outcomes, RAAS blockers can be reinstituted after successful management of hyperkalemia and preventive measures are in place [106]. Potassium elimination can be increased via the kidneys and the colon. These decisions are based on serum K,

	Sodium Polystyrene Sulfonate [108, 113]	**Patiromer [109]**	**Sodium Zirconium Cyclosilicate [107, 110, 111, 114, 115]**
FDA Approval	1958	2015	2018
Chemical Structure	Organic polymer/ resin	Cross-linked polymer	Inorganic microporous compound
Sodium Content	100 mg/g	None	80 mg/g
Mechanism of Action	Sodium/ Potassium exchange, nonspecifically binds potassium, magnesium, calcium	Calcium/ Potassium exchange, also binds magnesium	Potassium exchanged with sodium and hydrogen
Onset of Action	Hours to Days	7 hours	1 hour
Amount of Potassium lowered	1 meq/L with 30 g	1 meq/L with 8.4 g	0.7 meq/L with 10 g
Adverse Effects	Intestinal necrosis	Hypomagnesemia, diarrhea, abdominal discomfort, flatulence	Nausea, vomiting, hypokalemia

Table 2.
Colonic potassium binders [107–112].

urine output, GFR, and other comorbidities. In a patient with reasonable GFR, loop and thiazide diuretics can be used.

Potassium binders are resins that can be administered enterally (**Table 2**). These resins bind to K in the colon and promote its elimination in the stool, and they are very effective. However, they require one to several hours for onset of action and an intact bowel function [10, 107, 116]. With life-threatening cardiac changes and low GFR, dialysis may be indicated. Hemodialysis is most efficacious in eliminating K and will remove 50–80 meq of K in a standard 4-hour session. Peritoneal dialysis can be tried, but it is slower in removal of K [117, 118].

6. Conclusion

Potassium is an integral intracellular cation. Any major disturbances in its homeostasis can be detrimental. Transcellular shifting represents an effective initial body response to such disturbances, with the kidney as the ultimate site for final regulation. Both hypo- and hyperkalemia are serious clinical problems and are associated with poor survival. Effective management includes correction of K, investigating underlying causes, and cardiac monitoring.

DRI Direct Renin Inhibitor, ACEI Angiotensin converting enzyme inhibitor, ARB Angiotensin Receptor Blocker, AR Angiotensin Receptor, CNI Calcineurin Inhibitor, NSAID Nonsteroidal anti-inflammatory drugs ENaC Epithelial Sodium Channel.

Author details

Sairah Sharif[1*] and Jie Tang[2]

1 Division of Kidney Diseases and Hypertension, Department of Medicine, Stony Brook University Hospital, New York, USA

2 Division of Kidney Diseases and Hypertension, Department of Medicine, Alpert Medical School of Brown University, Rhode Island, USA

*Address all correspondence to: sairah.sharif1@gmail.com

References

[1] Malnic G et al. Regulation of K+ excretion. Seldin and Giebisch's The Kidney. 2013:1659-1715. DOI: 10.1016/b978-0-12-381462-3.00049-5

[2] Mount DB, Zandi-Nejad K. Disorders of potassium balance. Brenner and Rector's The Kidney. 2012:640-688. DOI: 10.1016/b978-1-4160-6193-9.10017-x

[3] Hoffman BF, Suckling EE. Effect of several cations on transmembrane potentials of cardiac muscle. The American Journal of Physiology. 1956;**186**:317-324

[4] Institute of Medicine. Dietary Reference Intakes: The Essential Guide to Nutrient Requirements. Washington, DC: National Academies Press; 2006

[5] Rose BD, Post TW. Clinical physiology of acid-base and electrolyte disorders. 5th ed. New York: McGraw-Hill, Medical Pub. Division; 2001

[6] Sterns RH, Cox M, Feig PU, Singer I. Internal potassium balance and the control of the plasma potassium concentration. Medicine. 1981;**60**:339-354

[7] Palmer BF, Clegg DJ. Physiology and pathophysiology of potassium homeostasis. Advances in Physiology Education. 2016;**40**:480-490

[8] Williams ME et al. Catecholamine modulation of rapid potassium shifts during exercise. The New England Journal of Medicine. 1985;**312**:823-827

[9] Aronson PS, Giebisch G. Effects of pH on potassium: New explanations for old observations. Journal of the American Socety of Nephrology. 2011;**22**:1981-1989

[10] Palmer BF. Potassium binders for hyperkalemia in chronic kidney disease—Diet, renin-angiotensin-aldosterone system inhibitor therapy, and hemodialysis. Mayo Clinic Proceedings. 2020;**95**:339-354

[11] Shannon M, Lovejoy FH Jr. Hypokalemia after theophylline intoxication. The effects of acute vs chronic poisoning. Archives of Internal Medicine. 1989;**149**:2725-2729

[12] Clemessy JL et al. Hypokalaemia related to acute chloroquine ingestion. Lancet. 1995;**346**:877-880

[13] Minella RA, Schulman DS. Fatal verapamil toxicity and hypokalemia. American Heart Journal. 1991;**121**: 1810-1812

[14] Pinter A, Dorian P, Newman D. Cesium-induced torsades de pointes. The New England Journal of Medicine. 2002;**346**:383-384

[15] Bhoelan BS, Stevering CH, van der Boog ATJ, van der, Heyden MAG. Barium toxicity and the role of the potassium inward rectifier current. Clinical Toxicology. 2014;**52**:584-593

[16] Lim S. Approach to hypokalemia. Acta Medica Indonesiana. 2007;**39**: 56-64

[17] Cannon SC. Voltage-sensor mutations in channelopathies of skeletal muscle. The Journal of Physiology. 2010;**588**:1887-1895

[18] Cavel-Greant D, Lehmann-Horn F, Jurkat-Rott K. The impact of permanent muscle weakness on quality of life in periodic paralysis: A survey of 66 patients. Acta Myologica. 2012;**31**:126-133

[19] Fontaine B. Periodic paralysis. Advances in Genetics. 2008;**63**:3-23

[20] Venance SL et al. The primary periodic paralyses: Diagnosis, pathogenesis and treatment. Brain. 2006;**129**:8-17

[21] Cannon SC. Channelopathies of skeletal muscle excitability. Comprehensive Physiology. 2015;**5**:761-790

[22] Lin S-H. Thyrotoxic periodic paralysis. Mayo Clinic Proceedings. 2005;**80**:99-105

[23] Zacchia M, Abategiovanni ML, Stratigis S, Capasso G. Potassium: From physiology to clinical implications. Kidney Diseases (Basel). 2016;**2**:72-79

[24] Giebisch G, Krapf R, Wagner C. Renal and extrarenal regulation of potassium. Kidney International. 2007;**72**:397-410

[25] Patel S, Rauf A, Khan H, Abu-Izneid T. Renin-angiotensin-aldosterone (RAAS): The ubiquitous system for homeostasis and pathologies. Biomedicine & Pharmacotherapy. 2017;**94**:317-325

[26] Wilson FH. Human hypertension caused by mutations in WNK kinases. Science. 2001;**293**:1107-1112

[27] Su X-T et al. Disruption of KCNJ10 (Kir4.1) stimulates the expression of ENaC in the collecting duct. American Journal of Physiology. Renal Physiology. 2016;**310**:F985-F993

[28] Wang M-X et al. Potassium intake modulates the thiazide-sensitive sodium-chloride cotransporter (NCC) activity via the Kir4.1 potassium channel. Kidney International. 2018;**93**:893-902

[29] Piechotta K, Lu J, Delpire E. Cation chloride cotransporters interact with the stress-related kinases Ste20-related proline-alanine-rich kinase (SPAK) and oxidative stress response 1 (OSR1). The Journal of Biological Chemistry. 2002;**277**:50812-50819

[30] Zhang C et al. KCNJ10 determines the expression of the apical Na-Cl cotransporter (NCC) in the early distal convoluted tubule (DCT1). Proceedings of the National Academy of Sciences of the United States of America. 2014;**111**:11864-11869

[31] Kamynina E, Staub O. Concerted action of ENaC, Nedd4-2, and Sgk1 in transepithelial Na(+) transport. American Journal of Physiology. Renal Physiology. 2002;**283**:F377-F387

[32] Arroyo JP, Ronzaud C, Lagnaz D, Staub O, Gamba G. Aldosterone paradox: Differential regulation of ion transport in distal nephron. Physiology. 2011;**26**: 115-123

[33] Wei Y, Zavilowitz B, Satlin LM, Wang W-H. Angiotensin II inhibits the ROMK-like small conductance K channel in renal cortical collecting duct during dietary potassium restriction. The Journal of Biological Chemistry. 2007;**282**:6455-6462

[34] Stockigt JR. Potassium metabolism. Anaesthesia and Intensive Care. 1977;**5**:317-325

[35] Musso CG. Potassium metabolism in patients with chronic kidney disease (CKD), part I: Patients not on dialysis (stages 3-4). International Urology and Nephrology. 2004;**36**:465-468

[36] Gennari FJ, John Gennari F, Segal AS. Hyperkalemia: An adaptive response in chronic renal insufficiency. Kidney International. 2002;**62**:1-9

[37] Sandle GI, McNicholas CM, Lomax RB. Potassium channels in colonic crypts. Lancet. 1994;**343**:23-25

[38] Lomax RB, Warhurst G, Sandle GI. Characteristics of two basolateral potassium channel populations in human colonic crypts. Gut. 1996;**38**:243-247

[39] Mathialahan T, Maclennan KA, Sandle LN, Verbeke C, Sandle GI. Enhanced large intestinal potassium permeability in end-stage renal disease. The Journal of Pathology. 2005;**206**:46-51

[40] Sandle GI, Gaiger E, Tapster S, Goodship TH. Evidence for large intestinal control of potassium homoeostasis in uraemic patients undergoing long-term dialysis. Clinical Science. 1987;**73**:247-252

[41] Lanham-New SA, Lanham-New SA, Lambert H, Frassetto L. Potassium. Advances in Nutrition. 2012;**3**:820-821

[42] Stone MS, Martyn L, Weaver CM. Potassium intake, bioavailability, hypertension, and glucose control. Nutrients. 2016;**8**:444

[43] Gumz ML, Rabinowitz L. Role of circadian rhythms in potassium homeostasis. Seminars in Nephrology. 2013;**33**:229-236

[44] Rabelink TJ, Koomans HA, Hené RJ, Dorhout Mees EJ. Early and late adjustment to potassium loading in humans. Kidney International. 1990;**38**:942-947

[45] Liamis G, Liberopoulos E, Barkas F, Elisaf M. Spurious electrolyte disorders: A diagnostic challenge for clinicians. American Journal of Nephrology. 2013;**38**:50-57

[46] Don BR, Sebastian A, Cheitlin M, Christiansen M, Schambelan M. Pseudohyperkalemia caused by fist clenching during phlebotomy. New England Journal of Medicine. 1990;**322**:1290-1292

[47] Colussi G, Cipriani D. Pseudohyperkalemia in extreme leukocytosis. American Journal of Nephrology. 1995;**15**:450-452

[48] Graber M, Subramani K, Corish D, Schwab A. Thrombocytosis elevates serum potassium. American Journal of Kidney Diseases. 1988;**12**:116-120

[49] Meenaghan M, Frederick Follett G, Brophy PJ. Temperature sensitivity of potassium flux into red blood cells in the familial pseudohyperkalaemia syndrome. Biochimica et Biophysica Acta (BBA) - Biomembranes. 1985;**821**:72-78

[50] Alani FS et al. Pseudohyperkalaemia associated with hereditary spherocytosis in four members of a family. Postgraduate Medicine Journal. 1994;**70**:749-751

[51] Meng QH, Krahn J. Reverse pseudohyperkalemia in heparin plasma samples from a patient with chronic lymphocytic leukemia. Clinical Biochemistry. 2011;**44**:728-730

[52] Naparstek Y, Gutman A. Case report: Spurious hypokalemia in myeloproliferative disorders. The American Journal of the Medical Sciences. 1984;**288**:175-177

[53] Sodi R, Davison AS, Holmes E, Hine TJ, Roberts NB. The phenomenon of seasonal pseudohypokalemia: Effects of ambient temperature, plasma glucose and role for sodium-potassium-exchanging-ATPase. Clinical Biochemistry. 2009;**42**:813-818

[54] Masters PW, Lawson N, Marenah CB, Maile LJ. High ambient temperature:

A spurious cause of hypokalaemia. BMJ. 1996;**312**:1652-1653

[55] Kardalas E et al. Hypokalemia: A clinical update. Endocrine Connections. 2018;7:R135-R146

[56] Dhondup T, Qian Q. Acid-base and electrolyte disorders in patients with and without chronic kidney disease: An update. Kidney Diseases (Basel). 2017;**3**:136-148

[57] Udensi UK, Tchounwou PB. Potassium homeostasis, oxidative stress, and human disease. International Journal of Clinical and Experimental Physiology. 2017;**4**:111-122

[58] Kovesdy CP et al. Serum potassium and adverse outcomes across the range of kidney function: A CKD Prognosis Consortium meta-analysis. European Heart Journal. 2018;**39**:1535-1542

[59] Collins AJ et al. Association of serum potassium with all-cause mortality in patients with and without heart failure, chronic kidney disease, and/or diabetes. American Journal of Nephrology. 2017;**46**:213-221

[60] Bartter FC, Pronove P, Gill JR Jr, Maccardle RC. Hyperplasia of the juxtaglomerular complex with hyperaldosteronism and hypokalemic alkalosis: A new syndrome. The American Journal of Medicine. 1962;**33**:811-828

[61] Simon DB et al. Bartter's syndrome, hypokalaemic alkalosis with hypercalciuria, is caused by mutations in the Na–K–2CI cotransporter NKCC2. Nature Genetics. 1996;**13**:183-188

[62] Watanabe S et al. Association between activating mutations of calcium-sensing receptor and Bartter's syndrome. Lancet. 2002;**360**:692-694

[63] Mendonca E, Stone R, Schurman S. Mutations in the chloride channel gene, CLCNKB, cause Bartter's syndrome type III. Nature. 1997;**17**(2):171-178

[64] Kurtz I. Molecular pathogenesis of Bartter's and Gitelman's syndromes. Kidney International. 1998;**54**:1396-1410

[65] Lifton RP et al. A chimaeric llβ-hydroxylase/aldosterone synthase gene causes glucocorticoid-remediable aldosteronism and human hypertension. Nature. 1992;**355**:262-265

[66] Tamura H et al. Liddle disease caused by a missense mutation of beta subunit of the epithelial sodium channel gene. The Journal of Clinical Investigation. 1996;**97**:1780-1784

[67] White PC. 11 -Hydroxysteroid dehydrogenase and the syndrome of apparent mineralocorticoid excess. Endocrine Reviews. 1997;**18**:135-156

[68] Edwards CR, Walker BR, Benediktsson R, Seckl JR. Congenital and acquired syndromes of apparent mineralocorticoid excess. The Journal of Steroid Biochemistry and Molecular Biology. 1993;**45**:1-5

[69] Ljubicic ML, Madsen A, Juul A, Almstrup K, Johannsen TH. The application of principal component analysis on clinical and biochemical parameters exemplified in children with congenital adrenal hyperplasia. Frontiers in Endocrinology. 2021;**12**:652888

[70] Huang C-L, Kuo E. Mechanism of hypokalemia in magnesium deficiency. Journal of the American Society of Nephrology. 2007;**18**:2649-2652

[71] Sundar S, Burma DP, Vaish SK. Digoxin toxicity and electrolytes: A correlative study. Acta Cardiologica. 1983;**38**:115-123

[72] Steiness E, Olesen KH. Cardiac arrhythmias induced by hypokalaemia and potassium loss during maintenance digoxin therapy. British Heart Journal. 1976;**38**:167-172

[73] Weiss JN, Qu Z, Shivkumar K. Electrophysiology of hypokalemia and hyperkalemia. Circulation. Arrhythmia and Electrophysiology. 2017;**10**(3): e004667

[74] Wang X, Han D, Li G. Electrocardiographic manifestations in severe hypokalemia. The Journal of International Medical Research. 2020;**48**:300060518811058

[75] Tizianello A et al. Renal ammoniagenesis in humans with chronic potassium depletion. Kidney International. 1991;**40**:772-778

[76] Yalamanchili HB, Calp-Inal S, Zhou XJ, Choudhury D. Hypokalemic nephropathy. Kidney International Reports. 2018;**3**:1482-1488

[77] Rastegar A. Attending rounds: Patient with hypokalemia and metabolic acidosis. Clinical Journal of the American Society of Nephrology. 2011;**6**:2516-2521

[78] Peter M, Sippell WG, Wernze H. Diagnosis and treatment of 17-hydroxylase deficiency. The Journal of Steroid Biochemistry and Molecular Biology. 1993;**45**:107-116

[79] Cerame BI, New MI. Hormonal hypertension in children: 11beta-hydroxylase deficiency and apparent mineralocorticoid excess. Journal of Pediatric Endocrinology & Metabolism. 2000;**13**:1537-1547

[80] Groeneveld JHM, Sijpkens YWJ, Lin S-H, Davids MR, Halperin ML. An approach to the patient with severe hypokalaemia: The potassium quiz. **QJM**. 2005;**98**:305-316

[81] Fleet JL et al. Validity of the international classification of diseases 10th revision code for hyperkalaemia in elderly patients at presentation to an emergency department and at hospital admission. BMJ Open. 2012;**2**:e002011

[82] Einhorn LM et al. The frequency of hyperkalemia and its significance in chronic kidney disease. Archives of Internal Medicine. 2009;**169**:1156-1162

[83] Sarafidis PA et al. Prevalence and factors associated with hyperkalemia in predialysis patients followed in a low-clearance clinic. Clinical Journal of the American Society of Nephrology. 2012;**7**:1234-1241

[84] Hayes J et al. Association of hypo- and hyperkalemia with disease progression and mortality in males with chronic kidney disease: The role of race. Nephron Clinical Practice. 2012;**120**: c8-c16

[85] Kaplan B et al. Frequency of hyperkalemia in recipients of simultaneous pancreas and kidney transplants with bladder drainage. Transplantation. 1996;**62**:1174-1175

[86] Adwaney A, Randall DW, Blunden MJ, Prowle JR, Kirwan CJ. Perioperative plasma-Lyte use reduces the incidence of renal replacement therapy and hyperkalaemia following renal transplantation when compared with 0.9% saline: A retrospective cohort study. Clinical Kidney Journal. 2017;**10**:838-844

[87] Gumz ML, Rabinowitz L, Wingo CS. An integrated view of potassium homeostasis. The New England Journal of Medicine. 2015;**373**:1787-1788

[88] Lehnhardt A, Kemper MJ. Pathogenesis, diagnosis and management of hyperkalemia. Pediatric Nephrology. 2011;**26**:377-384

[89] Preston RA, Afshartous D, Rodco R, Alonso AB, Garg D. Evidence for a gastrointestinal–renal kaliuretic signaling axis in humans. Kidney International. 2015;**88**:1383-1391

[90] Speiser PW et al. Congenital adrenal hyperplasia due to steroid 21-hydroxylase deficiency: An Endocrine Society clinical practice guideline. The Journal of Clinical Endocrinology and Metabolism. 2010;**95**:4133-4160

[91] Hui E et al. The clinical significance of aldosterone synthase deficiency: Report of a novel mutation in the CYP11B2 gene. BMC Endocrine Disorders. 2014;**14**:29

[92] Allen AM, Zhuo J, Mendelsohn FA. Localization of angiotensin AT1 and AT2 receptors. Journal of the American Society of Nephrology. 1999;(**10 Suppl. 11**):S23-S29

[93] Allen AM, Zhuo J, Mendelsohn FA. Localization and function of angiotensin AT1 receptors. American Journal of Hypertension. 2000;**13**:31S-38S

[94] Heering PJ et al. Aldosterone resistance in kidney transplantation is in part induced by a down-regulation of mineralocorticoid receptor expression. Clinical Transplantation. 2004;**18**: 186-192

[95] Kim HC et al. Primary immunosuppression with tacrolimus in kidney transplantation: Three-year follow-up in a single center. Transplantation Proceedings. 2004;**36**:2082-2083

[96] Hoorn EJ, Walsh SB, McCormick JA, Zietse R. Pathogenesis of calcineurin inhibitor–induced hypertension. Journal of Nephrology. 2012;**25**(3):269

[97] Slovis C, Jenkins R. ABC of clinical electrocardiography: Conditions not primarily affecting the heart. BMJ. 2002;**324**:1320-1323

[98] Levis JT. ECG diagnosis: Hyperkalemia. The Permanente Journal. 2013;**17**:69

[99] El-Sherif N, Turitto G. Electrolyte disorders and arrhythmogenesis. Cardiology Journal. 2011;**18**:233-245

[100] Noble K, Isles C. Hyperkalaemia causing profound bradycardia. Heart. 2006;**92**:1063

[101] Diercks DB, Shumaik GM, Harrigan RA, Brady WJ, Chan TC. Electrocardiographic manifestations: Electrolyte abnormalities. The Journal of Emergency Medicine. 2004;**27**:153-160

[102] Pochineni V, Rondon-Berrios H. Electrolyte and acid-base disorders in the renal transplant recipient. Frontiers in Medicine. 2018;**5**:261

[103] Ayach T, Nappo RW, Paugh-Miller JL, Ross EA. Life-threatening hyperkalemia in a patient with normal renal function. Clinical Kidney Journal. 2014;**7**:49-52

[104] John SK, Rangan Y, Block CA, Koff MD. Life-threatening hyperkalemia from nutritional supplements: Uncommon or undiagnosed? The American Journal of Emergency Medicine. 2011;**29**:1237.e1-1237.e2

[105] Chen CM, Gettes LS, Katzung BG. Effect of lidocaine and quinidine on steady-state characteristics and recovery

kinetics of (dV/dt)max in guinea pig ventricular myocardium. Circulation Research. Jul 1975;**37**(1):20-29

[106] Hou FF et al. Efficacy and safety of benazepril for advanced chronic renal insufficiency. The New England Journal of Medicine. 2006;**354**:131-140

[107] Kosiborod M et al. Effect of sodium zirconium cyclosilicate on potassium lowering for 28 days among outpatients with hyperkalemia: The HARMONIZE randomized clinical trial. JAMA. 2014;**312**:2223-2233

[108] Sterns RH, Rojas M, Bernstein P, Chennupati S. Ion-exchange resins for the treatment of hyperkalemia: Are they safe and effective? Journal of the American Society of Nephrology. 2010;**21**:733-735

[109] Weir MR et al. Patiromer in patients with kidney disease and hyperkalemia receiving RAAS inhibitors. The New England Journal of Medicine. 2015;**372**:211-221

[110] Lokelma approved in the US for the treatment of adults with hyperkalaemia. 2018. Available from: https://www.astrazeneca.com/media-centre/press-releases/2018/lokelma-approved-in-the-us-for-the-treatment-of-adults-with-hyperkalaemia-21052018.html

[111] Stavros F, Yang A, Leon A, Nuttall M, Rasmussen HS, Bansal V, et al. editors. In vitro ion exchange capacity and selectivity of ZS-9, a novel, selective cation trap for the treatment of hyperkalemia. In: Characterization of Structure and Function of ZS-9, a K^+ Selective Ion Trap. American Journal of Kidney Diseases. 2014;**63**:A115

[112] Ash SR, Singh B, Lavin PT, Stavros F, Rasmussen HS. A phase 2 study on the treatment of hyperkalemia in patients with chronic kidney disease suggests that the selective potassium trap, ZS-9, is safe and efficient. Kidney International. 2015;**88**:404-411

[113] Mikrut M, Brockmiller-Sell H. Sodium polystyrene sulfonate dosing guidelines for the treatment of adult hyperkalemia. Hospital Pharmacy. 2004;**39**:765-771

[114] Stavros F, Yang A, Leon A, Nuttall M, Rasmussen HS. Characterization of structure and function of ZS-9, a K+ selective ion trap. PLoS One. 2014;**9**:e114686

[115] Packham DK et al. Sodium zirconium cyclosilicate in hyperkalemia. The New England Journal of Medicine. 2015;**372**:222-231

[116] Rosano GMC, Spoletini I, Agewall S. Pharmacology of new treatments for hyperkalaemia: Patiromer and sodium zirconium cyclosilicate. European Heart Journal Supplements: Journal of the European Society of Cardiology. 2019;**21**:A28-A33

[117] Roseman DA, Schechter-Perkins EM, Bhatia JS. Treatment of life-threatening hyperkalemia with peritoneal dialysis in the ED. The American Journal of Emergency Medicine. 2015;**33**:473.e3-473.e5

[118] Ilabaca-Avendaño MB, Yarza-Solorzáno G, Rodriguez-Valenzuela J, Arcinas-Fausto G, Ramírez-Hernandez V, Hernández-Hernández DA, et al. Automated peritoneal dialysis as a lifesaving therapy in an emergency room: Report of four cases. Kidney International Supplements. Apr 2008;(108):S173-S176. DOI: 10.1038/sj.ki.5002620. PMID: 18379542

Section 3

Potassium in Heart and Cancer

Chapter 4

Potassium and Cardiac Surgery

Shawn Kant, Frank W. Sellke and Jun Feng

Abstract

Potassium homeostasis affects cardiac rhythm and contractility, along with vascular reactivity and vascular smooth muscle proliferation. This chapter will focus on potassium dynamics during and after cardiac surgery involving cardioplegic arrest and cardiopulmonary bypass (CPB). Hyperkalemic, hypothermic solutions are frequently used to induce cardioplegic arrest and protect the heart during cardiac surgery involving CPB. Common consequences of hyperkalemic cardioplegic arrest and reperfusion include microvascular dysfunction involving several organ systems and myocardial dysfunction. Immediately after CPB, blood potassium levels often drop precipitously due to a variety of factors, including CPB -induced electrolyte depletion and frequent, long-term administration of insulin during and after surgery. Meanwhile, some patients with pre-existing kidney dysfunction may experience post-operative hyperkalemia following cardioplegia. Any degree of postoperative hyper/hypokalemia significantly elevates the risk of cardiac arrythmias and subsequent myocardial failure. Therefore, proper management of blood potassium levels during and after cardioplegia/CPB is crucial for optimizing patient outcomes following cardiac surgery.

Keywords: potassium, hypokalemia, hyperkalemia, low potassium, high potassium, hyperkalemic cardioplegia, cardiopulmonary bypass, cardiac surgery, open heart surgery, postoperative arrhythmias, perioperative arrhythmias

1. Introduction

Intracellular and blood potassium levels have crucial effects on cardiovascular system homeostasis. At the most fundamental level, the potassium concentration gradient across cardiac muscle cell (cardiomyocyte) cell membranes is a chief determinant of cardiomyocyte resting membrane potentials. Indeed, disruptions to this concentration gradient (e.g. via increasing or decreasing extracellular blood potassium levels) can lead to altered cardiomyocyte contractility and excitability. Potassium is also vasoactive, with different effects at different extracellular concentrations. At low (5-8 mM) to moderate (8-16 mM) extracellular levels, potassium relaxes the smooth muscle in blood vessel walls by promoting hyperpolarization of vascular smooth muscle. However, at higher levels (16-25 mM and above) (e.g. cardioplegic concentrations), potassium promotes vasoconstriction by facilitating depolarization. Moreover, potassium is released by vascular endothelial cells in response to various chemical mediators and shear stress, thereby contributing to the action of endothelium-derived hyperpolarizing factor [1]. For all of these reasons and more, keeping track of daily potassium intake is often recommended as a lifestyle modification for chronic cardiovascular diseases such as hypertension.

Harnessing the pivotal role of potassium in cardiovascular physiology has proved quite useful for cardiovascular surgery, namely in the form of hyperkalemic (high potassium) cardioplegia. Indeed, throughout the past several decades, a large body of research has testified to the ability of externally administered hyperkalemic solutions to arrest cardiac contractility [2]. This, in conjunction with the development of cardiopulmonary bypass (CPB, also known as the "heart-lung machine"), revolutionized cardiac surgery [3]. These days, many highly invasive procedures like coronary artery bypass grafting are routine with minimal risk of postoperative mortality.

However, hyperkalemic cardioplegia is not without its consequences. Hyperkalemic cardioplegia and reperfusion following CPB have been associated with perioperative and postoperative tissue damage and microvascular dysfunction across several different vascular beds. Moreover, hyperkalemic cardioplegia is also associated with postoperative myocardial dysfunction and reduced cardiac output. Furthermore, blood potassium abnormalities after hyperkalemic cardioplegia-reperfusion, chiefly hypokalemia (but also hyperkalemia, to a lesser degree) are common postoperative challenges in the cardiac ICU. Both abnormalities significantly elevate the risk of arrythmias and, if not managed properly, cardiac arrest and sudden death.

This chapter will discuss the basics of potassium cardioplegia with an emphasis on clinical relevance, beginning with a brief history. Subsequent sections will elaborate on the basic physiology, before considering several perioperative and postoperative adverse effects of hyperkalemic cardioplegia. When possible, information about treatment and clinical management is included. The chapter will conclude with a brief mention of up-and-coming alternatives to hyperkalemic cardioplegia.

2. Potassium and cardiac surgery

2.1 Brief history of potassium cardioplegia

As early as the late 1800s, physiologists were starting to become aware of the ability of potassium compounds to arrest cardiac contractility, beginning with individuals like Sidney Ringer who observed that potassium chloride froze the heart in diastole and calcium stimulated the heart during systole [2]. Moving into the start of the 20th century, further investigations revealed associations between high serum potassium and cardiac arrest following ventricular fibrillation; studies also revealed associations between cardioplegia and restoration of sinus rhythm following coronary artery administration of potassium chloride solution and subsequent washout [2]. However, in most of these cardioplegic experiments (often conducted in dogs), refractory ventricular fibrillation and post-procedure reperfusion damage to the myocardium limited discussion of the clinical usefulness of these findings.

During the 1950s, British physician Dennis Melrose hypothesized that the problem with potassium chloride cardioplegia was chloride; therefore, he created a cardioplegic solution using potassium citrate, and tested it on a canine model of cardiopulmonary bypass [4]. Injection of the "Melrose solution", of potassium citrate plus warm oxygenated whole blood in a 9:1 blood:potassium ratio, into the aortic roots of hypothermic dogs, produced near-immediate cardiac arrest. Reperfusion and washout of cardioplegic solution resulted in restoration of heart function to pre-procedure levels [2]. Within a few years, the Melrose group successfully induced potassium citrate cardioplegia in humans.

Unfortunately, future studies would reveal that in many cases, the Melrose potassium citrate solution still produced post-cardioplegia ventricular fibrillation and myocardial dysfunction [5]. This led to a general pause in clinical application of potassium cardioplegia between the 1960s and early 1980s, in favor of other options mostly involving induction of hypothermic cardiac arrest, which turned out to be no better with respect to postoperative damage than the Melrose solution.

Eventually, research into techniques for potassium cardioplegia would pick up again, and the result would be development of novel solutions for cardioplegia and intraoperative organ preservation. Numerous studies in animal models have validated the principles of diastolic cardiac arrest due to depolarizing potassium cardioplegia [2, 3, 6–10]. In addition, invention and refinement of heart-lung machines to accompany cardioplegia in the operative room (CPB) opened many new possibilities for cardiac surgery. Today, potassium cardioplegia is an integral tool for cardiac surgeons performing a variety of highly invasive procedures such as coronary artery bypass grafting and aortic valve replacements.

2.2 Types and techniques of potassium cardioplegia

Despite variability in composition, delivery, and temperature, most cardioplegic solutions in use today involve some level of potassium chloride as the main inducer of cardiac arrest, along with ions such as magnesium, low-dose calcium and bicarbonate, the latter of which is particularly important for controlling solution pH [6]. The "original" hyperkalemic cardioplegic solution was the Melrose formula of the 1950s that was discussed earlier, consisting of potassium citrate and warm blood in a 9:1 blood:potassium ratio. However, due to the high incidence of postoperative complications including ventricular fibrillation, this solution is no longer in major clinical use.

2.2.1 Crystalloid vs. blood

In general, cardioplegic solutions fall under two broad umbrellas: crystalloid vs. blood, and warm vs. cold (**Table 1**). Two crystalloid cardioplegic solutions worth noting are the Custodiol (also known as Bretschneider) and St. Thomas solutions [7]. The St. Thomas solution, introduced first by Hearse and colleagues in 1975, is an example of a short acting cardioplegic solution involving potassium chloride concentrations between 10 and 30 mM [8]. In general, the St. Thomas solution requires repeat dosing, roughly every 20 minutes, to sustain cardioplegia for long durations [7, 9]. Furthermore, myocardial acidosis has been noted between doses of St. Thomas solution [10].

In contrast, the Custodiol solution is a form of long acting, single dose cardioplegia consisting primarily of potassium chloride, sodium chloride, and magnesium sulfate as the chief electrolytes [11]. Additional components of the Custodiol solution include tryptophan (membrane stabilization) and histidine buffer (to maintain pH and buffer against byproducts of anaerobic glycolysis that build up during cardioplegia). Curiously, the relatively low levels of potassium (9 mM) and sodium (15 mM) in Custodiol appear to induce cardioplegia through a form of hyperpolarized arrest as opposed to depolarized arrest, unlike most other potassium cardioplegic solutions that have potassium concentrations in the range of 16-36 mM and sodium concentrations in the range of 10-110 mM (see **Table 1** for detailed solution ion concentrations).

	St. Thomas Cardioplegia	Custodiol Cardioplegia	Del Nido Cardioplegia	Buckberg Cardioplegia	Warm Calafiore Cardioplegia (one variant)
K+	16 mM	9 mM	26 mM	Cold induction: 36 mM Maintenance: 36 mM Reperfusion: 15 mM	18–20 mM for inducing arrest, repeat delivery every 20 min with decreasing K concentrations
Ca	1.2 mM	0.015 mM			1.3 mM
Mg	16 mM	4 mM	2 g of 50% magnesium sulfate		15.5 mM
Na	110 mM	15 mM			
NaHCO3	10 mM		13 mM		
Other Components		18 mM Histidine hydrochloride 18 mM histidine 2 mM tryptophan 30 mM mannitol 1 mM potassium hydrogen 2-ketoglutarate	13 mL of 1% lidocaine 3.2 g/L of 20% mannitol	*Cold Induction:* 392 mL 5% dextrose, 50 mL 0.3 M tromethamine, 30 mL citrate–phosphate-2-dextrose *Maintenance*: 798 mL 5% dextrose, 123 mL 0.3 M tromethamine, 61 mL citrate–phosphate-2-dextrose *Reperfusion*: 26 mL 50% dextrose; 56 mL 0.3 M tromethamine; 113 mL citrate–phosphate-2-dextrose 62.5 mL glutamate/aspartate	500 mL 5% dextrose 4 mM tris(hydroxymethyl) aminomethane Core body temperature maintained at 37 degrees Celsius
Blood vs. Crystalloid	Crystalloid	Crystalloid	4:1 crystalloid: blood ratio	4:1 crystalloid: blood ratio	Normothermic blood

Table 1.
Composition of common potassium-based cardioplegic solutions.

The general rationale for blood-based cardioplegia has centered on the theory that cardioplegic solutions containing blood are more "physiologic" than crystalloid solutions. For example, blood can support aerobic respiration and may be able to preserve normal myocardial metabolism during surgery. Therefore, blood cardioplegia may reduce the negative consequences of prolonged ischemia during CPB [11]. However, insufficient evidence exists currently to verify that hypothesis, and so any purported advantages of blood over crystalloid cardioplegia are for the time being mainly speculative.

Three hyperkalemic cardioplegic solutions in clinical use that contain blood are the Del Nido, Buckberg, and Calafiore solutions. The Del Nido solution uses a crystalloid:blood ratio of 4:1, and like the Custodiol solution is a long-acting cardioplegic solution, with one dose of 20 ml/kg providing myocardial protection for up to 60–90 minutes [7, 12]. Chief ionic ingredients include potassium chloride for rapid depolarized arrest, sodium bicarbonate to scavenge protons and buffer intracellular pH, and magnesium to block calcium channels and prevent intracellular calcium accumulation during cardioplegic arrest, thereby promoting postoperative myocardial recovery [12, 13]. Lidocaine in the Del Nido solution acts as a sodium channel blocker to mitigate against the sodium "window current" and reduce intracellular sodium accumulation [14].

Buckberg's cardioplegia is a dextrose and saline-based solution that, similar to the Del Nido solution, consists of a crystalloid:blood ratio of 4:1 [15]. Other components include potassium chloride as the primary depolarizing agent, a tromethamine buffer, and citrate phosphate double dextrose to serve as a calcium chelator. However, unlike the Del Nido solution, Buckberg cardioplegia must be given as three separate formulations, some of which must be administered in multiple doses [15]. First, an induction solution stops the heart, and additional infusions of induction solution must be given every 15 to 20 minutes throughout the procedure. Second, a maintenance solution must be administered to sustain cardiac arrest and provide oxygen and nutrients to the cardiomyocytes. Finally, a reperfusion solution containing glutamate and aspartate is administered prior to removal of the aortic cross clamp to provide the heart with nutrients prior to restarting myocardial contractions.

Calafiore cardioplegia differs from Buckberg and Del Nido in that blood forms the sole foundation of Calafiore cardioplegic solution [16]. Indeed, the original rational proposed by Calafiore et al. was that blood alone, without any crystalloid component, contained everything necessary to prevent ischemia–reperfusion damage. Therefore, simply administering a cardioplegic solution consisting of blood plus extra potassium would be enough to safely stop and later, restart the heart [16]. Moreover, unlike most other forms of cardioplegia in use, the original Calafiore solution was normothermic throughout administration; however, some subsequent variations of Calafiore cardioplegia have used cold blood [16, 17].

2.2.2 Warm vs. cold

Most current methods for administering cardioplegic solutions involve cold cardioplegia, most often cold crystalloid solutions delivered after reducing core body temperature to hypothermic levels [18]. For example, the induction and maintenance solutions for Buckberg cardioplegia are delivered at 4 degrees Celsius after cooling core temperature to below 30 degrees Celsius, with reperfusion solution delivered at 37 degrees Celsius [15]. Similarly, del Nido and Custodiol cardioplegia are often given at 4 degrees Celsius after induction of systemic hypothermia [15, 19].

This practice stems from experimental evidence suggesting that mild hypothermia can protect the myocardium from ischemic damage during cardioplegia [20]. Hypothermia reduces the basal metabolic rate of the heart, which in turn reduces oxygen consumption—an effect augmented by potassium-induced arrest during hyperkalemic cardioplegia [21]. A variety of potential mechanisms may be at play. In animal models of cardiac arrest, mild hypothermia (32–35 degrees Celsius) has been shown to reduce post-arrest infarct size, possibly through various signal transduction pathways, such as Akt and mTOR signaling, both of which are altered during the course of hypothermia [20]. Another potential cardioprotective mechanism of hypothermia may be reduced phosphorylation of various mitogen activated protein kinases (MAPK) like ERK1/2 that normally activate pro-inflammatory mediators like COX-2 (arachidonic acid metabolism) [18]. In general, many details concerning mechanisms of hypothermic myocardial protection during cardioplegia remain to be elucidated.

However, cold hyperkalemic cardioplegia may also inhibit myocardial enzymes that are important for the metabolic and functional recovery of the heart after surgery [22, 23]. Moreover, sustained systemic hypothermia (especially at temperatures below 20 degrees Celsius) during cardiac surgery has also been associated with ventricular fibrillation after rewarming [21]. Given these negative consequences, an increasing amount of attention has been given to the possibility of warm hyperkalemic cardioplegia, primarily warm blood hyperkalemic cardioplegia. Unlike cold hyperkalemic cardioplegic solutions, warm cardioplegic solution is typically administered at between 30 and 35 degrees Celsius under normothermic, as opposed to hypothermic, CPB [24]. Potential advantages of warm blood hyperkalemic cardioplegia over cold crystalloid may include improved myocardial restoration, reduced intracellular swelling, improved membrane stabilization, and reduced hypoxic red blood cell deformation [25].

Of course, warm hyperkalemic cardioplegia is not without its own consequences. Some studies have reported increased likelihoods of perioperative strokes and encephalopathy [26]. Moreover, warm hyperkalemic cardioplegia may contribute to vasodilation during cardiopulmonary bypass, requiring increased use of alpha agonists during operation to maintain stable arterial perfusion pressures [25]. There are also several variations of warm cardioplegia; one common technical variant is "hot shot" cardioplegia, which involves warm induction and subsequent cold cardioplegia, followed by a warm reperfusion [27].

Comparing the effectiveness of warm vs. cold hyperkalemic cardioplegia remains an inconclusive subject of intense debate. A meta-analysis by Fan et al., reported no differences between length of stay, stroke incidence, and atrial fibrillation between patients undergoing warm vs. cold cardioplegia [28]. However, warm cardioplegia correlated with better postoperative cardiac indices and lower peak creatine kinase MB concentrations than cold cardioplegia [28]. The latter findings, along with reduced postoperative cardiac troponin levels, have been replicated in other studies [29, 30]. Meanwhile, other studies comparing warm blood and cold crystalloid hyperkalemic cardioplegia do not show significant differences with respect to perioperative myocardial infarction and low cardiac output syndrome [31].

2.2.3 Anterograde vs. retrograde

In general, administration of hyperkalemic cardioplegic solution can be done in either retrograde or anterograde fashion. Prior to both, IV heparin is administered, and the patient's core body temperature is lowered to hypothermic levels, after which the aortic cross-clamp is placed and cardiopulmonary bypass is initiated [7]. Anterograde cardioplegia refers to delivering cardioplegic solution through a cannula inserted just proximal to the aortic cross-clamp. From there, the solution can flow into the left and right coronary arteries that supply the myocardium [32]. With anterograde cardioplegia, arrest usually occurs within 30 to 60 seconds. Retrograde cardioplegia may be considered in patients with complications such as severe coronary artery damage (e.g. severe stenosis) or aortic valve damage. Unlike anterograde administration, in retrograde administration the cardioplegia catheter is inserted into the coronary sinus from the right atrium, and solution is injected at a lower pressure (given the lower tolerance of the coronary sinus walls to turbulent flow) to avoid coronary sinus perforation [32].

2.3 Physiology of potassium cardioplegia during cardiac surgery

2.3.1 Physiology of cardiac muscle contraction

Under physiological circumstances, the cardiomyocyte resting membrane potential is largely determined by two key factors: action of the sodium-potassium ATPase, and the high resting permeability of cardiomyocyte cell membranes to potassium [33]. First, the sodium-potassium ATPase hydrolyzes ATP to continuously pump potassium into the cell and sodium out of the cell, with a relative ratio of 3Na out/2 K in per molecule of ATP. Because it is the primary ion pump active while the cell is at rest, the sodium-potassium ATPase plays a critical role in generating the characteristic sodium and potassium electrochemical gradients across the cardiomyocyte cell membrane (high potassium and low sodium inside the cell relative to out). Second, at rest the cardiomyocyte cell membrane is most permeable to potassium while being relatively impermeable to other ions. This results in a resting membrane potential for cardiomyocytes that is close to the Nernst equilibrium potential for potassium, roughly −85 to -90 mV.

During cardiac muscle contraction, sinoatrial node stimulation induces a transient increase in the resting membrane potential of cardiomyocytes, which in turn opens voltage-gated sodium channels once the membrane potential surmounts -65 mV. Due to the high inward ion driving force on sodium (based on the considerable difference between the Nernst potential for sodium and the resting membrane potential), sodium ions flow through the sodium channels into the cardiomyocyte and further depolarize the cell until it reaches about 20 mV. At this point, sodium channels inactivate and L-type voltage gated calcium channels take over the maintenance of the action potential, allowing influx of calcium ions and producing the classic plateau depolarization of cardiac ventricular action potentials. Eventually, as calcium channels close and membrane potential begins to dip, delayed rectifier potassium channels open and restore membrane potential to the resting state. By this point, enough calcium has entered the cardiomyocyte to promote calcium-induced calcium release from intracellular calcium stores in the cardiomyocyte sarcoplasmic reticula, allowing muscle contraction to occur.

2.3.2 Physiology of potassium cardioplegia

Extracellular hyperkalemia is the core principle underpinning most warm blood and cold crystalloid cardioplegic solutions. Essentially, administration of hyperkalemic solution takes advantage of the pivotal role of the potassium electrochemical gradient in determining cardiomyocyte resting membrane potential in order to elevate the resting membrane potential to a less negative value than typical baseline level. For example, physiologic extracellular potassium levels are often in the range of 3.5–5 mM, producing a resting membrane potential around -85 mV. During cardiac surgery involving cardioplegia, hyperkalemic solutions often raise extracellular potassium to the range of 10-40 mM (often midway in this range, around the 25 mM level), elevating cardiomyocyte resting membrane potentials to anywhere between −65 to -40 mV [34]. Arresting cardiomyocytes at this new range of elevated membrane potentials promotes fast sodium channel inactivation, thereby blocking myocardial action potential conduction. It also blocks repolarization, which is why hyperkalemic cardioplegia induces what is called "depolarized arrest." Finally, it is important to note that cardioplegic arrest also significantly reduces cardiomyocyte oxygen consumption in a manner reminiscent of how severe ischemia depletes cellular ATP reserves [33].

2.4 Side effects of high potassium cardioplegia

2.4.1 Myocardial calcium loading

Despite its clinical usefulness in reversibly arresting the heart during cardiac surgery, sustained depolarized hyperkalemic cardioplegia is not without some negative perioperative consequences. First, while most voltage-gated "fast" sodium channels are inactivated at membrane potentials above -50 mV (a frequent target cardiomyocyte membrane potential for potassium cardioplegia), resulting in generally poor membrane sodium conductance, not *all* sodium channels are inactivated. Moreover, during hyperkalemic cardioplegia the ion driving force on sodium is still quite high, even at the new depolarized cell membrane potentials. Ultimately, this situation produces a small but significant sodium influx into cardiomyocytes through the small fraction of sodium channels that remain open during potassium cardioplegia, a phenomenon known as the sodium "window current" [35].

Similarly, ATP depletion and reduced myocardial oxygen consumption during hyperkalemic cardioplegia leads to myocardial ischemia. Ischemia forces myocardial cells to resort to anaerobic glycolysis for energy production, which generates lactate as a byproduct. Increasing lactate levels in cardiomyocytes produces a metabolic acidosis and promotes increased activity of the H+/Na antiporter to move protons out of the cells at the expense of bringing in more sodium [36]. Finally, the combination of high extracellular potassium, intracellular acidosis, and hypothermia due to cold cardioplegic solution inhibits action of the sodium-potassium ATPase, which further facilitates the buildup of intracellular sodium [34].

Note that -50 mV is also in the vicinity of the reversal potential of the sodium/calcium exchanger [37, 38]. Under normal circumstances, the sodium/calcium exchanger moves 3 Na in for every 1 Ca moved out of the cell. However, due to the sodium window current and depolarized arrest in hyperkalemic cardioplegia, the sodium/calcium exchanger eventually begins operating in reverse, moving 3 Na out for every 1 Ca in, producing a so-called calcium "window current." Moreover, if the hyperkalemic cardioplegic solution holds cardiomyocyte membrane potentials above -50 mV, e.g. at around -40 mV, then voltage-gated slow calcium channels will begin to activate, causing further calcium influx [39]. All of these reasons help explain why many hyperkalemic cardioplegic solutions in clinical practice are also hypocalcemic relative to physiological extracellular calcium levels (or contain calcium channel blockers), to attempt to mitigate the severity of myocardial calcium loading [34].

Cytosolic calcium loading during hyperkalemic cardioplegia contributes to cardiomyocyte damage through several mechanisms [40]. Enhanced activation of calcium dependent proteases and lipases (e.g. phospholipases) contributes to plasma membrane phospholipid degradation, ultrastructural changes in the sarcolemmal membrane, and accumulation of pathological catabolic byproducts. Enhanced activation of calcium-dependent ATPases accelerates depletion of intracellular ATP stores that have already been lowered following hypothermic arrest. This further perturbs cardiomyocyte sarcolemmal calcium transport channels that rely on ATP to maintain intracellular calcium homeostasis. Moreover, hypoxia during hyperkalemic cardioplegia increases mitochondrial calcium uptake via reversal of mitochondrial sodium/calcium exchangers in a manner akin to reversal of cardiomyocyte cell membrane sodium/calcium exchangers [41].

Mitochondria can only endure so much calcium uptake before the onset of irreversible damage. Indeed, following reperfusion after hyperkalemic cardioplegia arrest, mitochondria exhibit increased oxygen free radical production and reduced superoxide dismutase activity, indicative of heightened oxidative stress [41]. Sustained oxidative stress can lead to opening of mitochondrial permeability transition pores (MPTP), which promote mitochondrial swelling and mitochondrial membrane rupture. An assortment of mitochondrial enyzmes and molecules, such as cytochrome c, leak out into the cytosol through the MPTPs [41]. Cytochrome c is implicated in intrinsic apoptotic pathways through activation of cytosolic caspases and subsequent formation of myocardial apoptosomes [41].

2.4.2 Myocardial apoptosis

Myocardial apoptosis during hyperkalemic cardioplegic ischemia–reperfusion merits further consideration for two major reasons. First, several studies have shown associations between hyperkalemic cardioplegic arrest and endothelial cell and cardiomyocyte apoptosis [42–44]. Second, several independent pathways of myocardial cell injury converge on apoptosis. Examples include mitochondrial oxidative stress and activation of an intrinsic apoptotic pathway (introduced earlier), or an extrinsic pathway driven by elevated humoral factors such as Fas or TNF-alpha acting on pro-apoptosis cell membrane receptors [44, 45]. Both intrinsic and extrinsic pathways converge upon a similar final common pathway that is chiefly regulated by two key protein groups: the Bcl-2 and cysteine protease caspase families [46, 47].

Within the Bcl-2 family, two proteins are particularly significant: Bcl-2 itself, and Bad. The former is anti-apoptotic while the latter is pro-apoptotic. Phosphorylation inhibits Bad, blocking it from inactivating Bcl-2 [48]. Farther downstream in apoptotic signaling, cleavage of caspase 3 and poly ADP-ribose polymerase (PARP) is essential for ensuring final progression towards apoptosis. Meanwhile, apoptosis may also proceed via a caspase-independent pathway involving release of the mitochondrial flavoprotein apoptosis-inducing factor (AIF) from the mitochondria into the cytosol through MPTPs [49, 50].

A possible framework for understanding myocardial apoptosis after hyperkalemic CPB is as follows [48]. Activation of the intrinsic (mitochondrial) pathway leads to increased Bad activation/decreased Bcl-2 activation, which initiates the caspase cascade. Activation of the extrinsic pathway bypasses Bcl-2/Bad to directly activate the caspase cascade. As more and more caspases become activated, eventually terminal caspases, such as caspase 3, will be cleaved, leading to PARP cleavage. By this point, apoptosis has been irreversibly induced; DNA fragmentation and cell death quickly follow. In contrast, AIF translocation from the mitochondria to the cytosol may directly activate downstream/terminal caspases, bypassing initial/intermediary constituents of the caspase cascade.

Studies have shown that caspase 3 cleavage and Bcl-2/Bad phosphorylation are significantly increased in myocardial tissue following hyperkalemic cold-blood cardioplegia and reperfusion, even as total protein levels do not change [48]. Meanwhile, myocardial AIF levels increase slightly, accompanied by a trend towards nuclear translocation, consistent with a model of AIF induced chromatin condensation and DNA fragmentation as a mechanism of cell injury [48]. Note that both pro-apoptotic (e.g. caspase 3) and anti-apoptotic (e.g. phosphorylated Bad) mediators are activated—nevertheless, given the downstream terminal position of caspase 3, the overall balance in myocardial cells appears to be tipped in favor of pro-apoptotic signaling.

Different formulations of hyperkalemic cardioplegia (e.g. cold crystalloid, warm blood, etc.) may exhibit differing degrees of myocardial protection and prevention of apoptosis. Indeed, evidence exists suggesting that cold blood hyperkalemic cardioplegia is superior to warm blood, warm crystalloid, and cold crystalloid cardioplegia, in terms of increased Bad phosphorylation and decreased caspase 3 activation [51]. Taken together, this combination of events appears to result in less apoptosis. In addition, these effects are associated with improved left ventricular function following cardioplegic arrest. However, this is not a universal finding in the literature. More work must be done to verify these conclusions and confirm if there truly is a definitive benefit to any one technique of hyperkalemic cardioplegia with respect to prevention of apoptosis.

2.4.3 Coronary vasomotor dysfunction

An extensive body of research has established that hyperkalemic solutions induce significant vasoconstriction when experimentally applied to coronary artery and aortic ring preparations [2]. Thus, it is no surprise that hyperkalemic cardioplegia induces significant functional changes to the microcirculation, especially the coronary circulation [52]. For example, a sizeable number of patients undergoing hyperkalemic cardioplegia experience coronary artery spasm [52].

Potassium can influence coronary vasoconstriction in several ways. Holding coronary vascular smooth muscle membrane potentials at sustained depolarization during hyperkalemic cardioplegia increases the likelihood of generating contractions [53]. Potassium may also act indirectly to cause vasospasm through action on the coronary endothelium. Indeed, endothelial vasoconstrictive and vasorelaxant factors govern homeostatic regulation of coronary vasomotor tone. These factors influence vascular smooth muscle through modulation of various cell membrane potassium channels, including calcium-activated potassium channels and ATP-activated potassium channels [54, 55]. Important endothelial-derived relaxing factors include nitric oxide, endothelial-derived hyperpolarizing factor (EDHF), and cyclooxygenase enzymes. Important endothelial-derived constricting factors include endothelin-1 and thromboxane A2.

Porcine models of hyperkalemic cardioplegia showed that hyperkalemia significantly attenuated EDHF-mediated relaxation in coronary artery preparations [56, 57]. Moreover, hyperkalemic vasoconstriction has also been linked with impaired nitric oxide release [58] and impaired acetylcholine-dependent vascular relaxation [59, 60]. Potential mechanisms at play may involve potassium-induced inhibition of G protein and non-G protein signal transduction pathways, increased reactive oxygen and nitrogen species generation, decreased activity of endothelial nitric oxide synthase, and increased arachidonic acid metabolism [2]. Curiously, hyperkalemic cardioplegia has also been associated with decreased responsiveness of human coronary arterioles to the endothelial vasoconstrictors endothelin-1 and thromboxane A2 [61, 62]. These findings testify to the complexity of mechanisms underpinning coronary vasomotor dysfunction following hyperkalemic cardioplegia, most of which remain to be elaborated.

2.4.4 Myocardial and coronary endothelial dysfunction

Despite its cardioprotective effects, hyperkalemic cardioplegia-reperfusion can exert detrimental effects on the myocardial and coronary endothelium, promoting endothelial dysfunction [63, 64]. One aspect of endothelial dysfunction—production of various endothelium-derived relaxing and contracting factors—was discussed earlier due to its relevance in coronary vasospasm. Other important features of endothelial dysfunction during hyperkalemic cardioplegic arrest include endothelial injury, inflammation, reactive oxygen species production, coagulation cascade dysfunction, and endothelial tight junction degradation [52, 65–67]. All these adverse effects may occur with potassium levels as low as 10 mM, well within the realm of most hyperkalemic cardioplegic solutions [2]. To elaborate, potassium concentrations of 30 mM in St. Thomas and Custodiol cardioplegic solutions proved considerably more damaging to the vascular endothelium than potassium concentrations of 20 mM, demonstrating the importance of strict potassium limits in hyperkalemic cardioplegic solutions [6].

A variety of structural changes to the vascular endothelium have been observed in experimental models of hyperkalemic cardioplegia. Key examples include endothelial intracellular vacuolization, membrane blebbing, adventitial fibrosis, and overall reduced viability [68, 69]. Furthermore, hyperkalemic cardioplegia promotes increased lipid uptake and cholesterol deposition in vascular intimae in primate models of post-graft venous atherosclerosis [70]. In addition, compromised endothelial adherens junctions during hyperkalemic cardioplegia mediate increased vascular permeability and tissue edema [67]. Indeed, animal models of cardioplegia/CPB show increased post-procedure VE cadherin, beta-catenin, and gamma-catenin fragments, all of which are important structural components of adherens junctions [71]. In humans, increased endothelial cadherin phosphorylation, and decreased overall beta-catenin levels, have been observed in atrial tissue following hyperkalemic cardioplegia/CPB [72].

Details of specific mechanisms underlying these endothelial disturbances remain largely unclear; however, many possibilities exist. For example, it is generally agreed that depolarization induced by hyperkalemic cardioplegia is a critical initiating step of the underlying pathophysiology [2]. Endothelial depolarization increases activation of neutrophils, inflammation, voltage sensitive NAPDH oxidases, and platelets [62, 63, 73, 74]. Inflammation and neutrophil activation often reinforce each other, as pro-inflammatory cytokines like IL-1, IL-6, and TNF-alpha further stimulate endothelial changes that promote neutrophil extravasation. NADPH oxidase catalyzes formation of important reactive oxygen species such as superoxide anions, which if left unchecked are severely cytotoxic. The amount of superoxide production during hyperkalemic cardioplegia has been linked to the extent of endothelial depolarization and translocation of the small G protein Rac from the cytosol to plasma membrane [75].

With respect to coagulation, potassium depolarization appears to have a direct stimulatory effect via enhancing ADP and collagen-induced platelet aggregation, along with an indirect effect through increased superoxide production [76, 77]. The latter appears to act through inhibition of endothelial NTPDases [78]. Membrane hyperpolarization reverses all these actions.

When left unchecked, sustained myocardial dysfunction following hyperkalemic cardioplegia-reperfusion may lead to myocardial stunning, a form of postoperative left ventricular dysfunction [1]. Myocardial stunning often manifests as markedly reduced cardiac output without obvious evidence of infarction or injury (e.g. no signs of elevated troponin or CKMB in blood). Like myocardial apoptosis, myocardial stunning represents another final common pathway of convergence for several different pathophysiological mechanisms of hyperkalemic cardioplegia, chiefly dysregulated free radical production, coagulation imbalances, and excessive catecholamine release [1]. However, unlike with apoptosis, in this scenario injury results from abnormal myocardial contractility as opposed to myocardial cell death.

2.5 Postoperative potassium abnormalities: physiology and management

Postoperative imbalances in a variety of different electrolytes, including calcium, magnesium, potassium, and phosphate, have been observed following cardioplegia/CPB. Here, we will focus on potassium, beginning with hypokalemia. Hypokalemia can be defined as a serum potassium level that is less than 3.5 mEq/L [78]. Postoperative hypokalemia is a common finding after cardiac surgery involving hyperkalemic cardioplegia and CPB, and manifests almost immediately after the patient is weaned off the bypass circuitry [79]. Hence IV potassium supplementation during cardioplegia is extremely important to mitigate against the most severe manifestations [80].

However, even with electrolyte supplementation in the operating room, CPB poses a high risk of post-procedure electrolyte depletion [81]. The pivotal role of potassium in normal cardiac contractility means that disturbances in potassium homeostasis significantly increase the risk of arrythmias and, in severe cases, sudden cardiac arrest. Indeed, arrythmias, especially atrial tachyarrhythmias (e.g. atrial fibrillation, atrial flutter) and, less frequently, ventricular arrhythmias, are a major source of morbidity and mortality following cardiac surgery [82, 83].

Specific mechanisms underpinning this phenomenon remain largely unclear; however, a variety of possibilities exist [78]. For example, poor oral intake of potassium-rich foods prior to cardiac surgery may contribute to enhanced depletion during surgery. In addition, prolonged preoperative use of digoxin, along with thiazide and loop diuretics may play a role. These agents may cause hypomagnesemia (low magnesium levels), which can contribute to extracellular potassium depletion. Under normal circumstances, intracellular magnesium binds to and blocks the pores of renal outer medullary potassium (ROMK) channels in the distal nephron, preventing outward flux of potassium into the renal tubular network [78]. Thus hypomagnesemia may remove this physiologic limiter, leading to increased renal clearance of potassium.

A hyperactive aldosterone response to stress may also be implicated, particularly in the context of congestive heart failure [78, 80]. Moreover, increased catecholamine (norepinephrine and epinephrine) release during cardiopulmonary bypass may facilitate hypokalemia given the influence of catecholamines on plasma potassium [84, 85]. Animal models have shown that elevated catecholamine levels can produce first, a transient hyperkalemia due to activation of hepatic calcium-dependent potassium channels by alpha adrenergic stimulation and second, a sustained hypokalemia by stimulation of skeletal muscle Na-K ATPase [86]. Such studies need to be replicated in humans undergoing cardiopulmonary bypass-hyperkalemic cardioplegia in order to verify the applicability of these putative mechanisms.

Because glucose is often given during cardioplegia, insulin may also be administered to minimize the chances of hyperglycemia. However, given that insulin acts as a regulator of potassium distribution between intracellular and extracellular fluid compartments by stimulating Na-K ATPase activity, it is possible that insulin administration during and after cardioplegia may contribute to potassium depletion [87]. Next, given that many cardioplegic solutions in current practice are cold hyperkalemic solutions, any potential impact of hypothermia on potassium homeostasis during cardiac surgery cannot be ignored. As with insulin, hypothermia has been linked to an intracellular shift of potassium away from the extracellular space through as-yet unelaborated mechanisms [88]. Finally, the CPB circuit itself has been shown to significantly dilute overall blood plasma protein concentrations, which may also affect plasma ion homeostasis [89].

In general, treatment of postoperative hypokalemia largely centers on administration of potassium chloride (KCl) solution to elevate extracellular potassium concentrations to physiologic levels. Indeed, in the case of pediatric cardiac ICU patients for whom enteral potassium supplementation is contraindicated, IV KCl administration is one of the only available tools for correcting hypokalemia [90]. For most patients, this proves sufficient to correct the imbalance and stave off the development of hypokalemia-induced arrhythmias. However, in a small minority, external KCl solution does not reverse the hypokalemia—and so in these patients, the chances of arrhythmias increase exponentially.

Although hypokalemia is the most common potassium electrolyte abnormality following hyperkalemic cardioplegia-CPB, postoperative hyperkalemia may occur under certain, albeit rarer, circumstances. In general, postoperative hyperkalemia is a concern mainly in patients with renal failure undergoing CPB, most likely due to renal tubular dysfunction [91]. Severe hyperkalemia may be treated with IV calcium gluconate, an insulin-dextrose regimen, and diuretics [92]. If a patient has end-stage renal disease, dialysis may be the best option to treat hyperkalemia, along with IV calcium to stabilize the myocardium and IV insulin to shift potassium into cells [93].

2.6 Alternatives to hyperkalemic cardioplegia

Hyperkalemic cardioplegia is by far the most widely used method of cardioplegia in current clinical practice. However, because of the numerous perioperative repercussions of hyperkalemic cardioplegia, a variety of attempts have been made to explore alternative approaches. Given that many adverse effects of hyperkalemic cardioplegia stem from its induction of depolarized arrest, one popular avenue of investigation has been the possibility of hyperpolarized arrest. Hyperpolarization is the natural resting state of cardiomyocytes, so in theory, arresting the heart at its baseline hyperpolarized state may better preserve physiological integrity. In isolated animal heart models, hyperpolarized arrest has been achieved via pharmacologic activation of ATP-sensitive potassium channels [94, 95]. Following reperfusion, this form of hyperpolarized arrest appeared to lead to improved postischemic functional recovery when compared to hearts protected with depolarized arrest.

Meanwhile, so-called "polarized arrest" has been proposed as another alternative to hyperkalemic cardioplegia. The core principle behind this concept is administration of sodium channel blockers, such as procaine in humans or tetrodotoxin in animal models [96]. Sodium channel blockade prevents depolarization-induced activation of calcium currents, which normally carry out the bulk of the cardiomyocyte action potential. Overall, in animal models, tetrodotoxin-induced polarized arrest reduces metabolic demands during ischemia, including myocardial oxygen consumption, more so than hyperkalemic cardioplegia [96]. Furthermore, polarized arrest may produce less significant postoperative ionic imbalances, with further protection provided by coincident administration of sodium/potassium/chloride transporter and sodium/proton exchanger inhibitors [96]. Nonetheless, more work needs to be done to verify the broader clinical applicability of these alternatives to hyperkalemic cardioplegia.

3. Conclusions

By taking advantage of the pivotal role of potassium in cardiomyocyte physiology, hyperkalemic cardioplegia has become an integral tool for cardiac surgery. From the early days of Dennis Melrose's simple potassium citrate solution to complex modern-day formulations such as the Del Nido and Buckberg media, approaches to developing and administering hyperkalemic cardioplegic solutions have evolved considerably, with a continuing focus on developing the most cardioprotective and least damaging solutions possible. While initial approaches to hyperkalemic cardioplegia revolved around hypothermic solutions, normothermic/"warm" solutions, along with blood as opposed to crystalloid-based solutions, are gaining momentum as potential alternatives to mitigate adverse perioperative consequences of cold hyperkalemic cardioplegia. Some of those consequences include myocardial calcium loading, myocardial apoptosis, coronary vasomotor dysfunction, myocardial endothelial dysfunction, and myocardial stunning. With any form of hyperkalemic cardioplegia, plasma potassium abnormalities following reperfusion, mainly postoperative hypokalemia, remain a persistent clinical concern. And while most patients respond well to IV KCl supplementation, some do not and proceed to develop fatal arrythmias, underscoring the need for further research to understand the mechanisms at play and develop new treatments. In the future, it is possible that other approaches such as hyperpolarized or polarized arrest may challenge the widespread use of depolarized hyperkalemic cardioplegic arrest. Nevertheless, for the time being, hyperkalemic cardioplegia remains dominant in cardiac surgery, and will likely continue to be so for some time to come.

Acknowledgements

The authors have no acknowledgements.

Conflict of interest

The authors declare no conflicts of interest.

Notes/thanks/other declarations

The authors have no other notes or declarations.

Author details

Shawn Kant, Frank W. Sellke and Jun Feng*
Division of Cardiothoracic Surgery, Department of Surgery, Rhode Island Hospital, Alpert Medical School of Brown University, Providence, Rhode Island, USA

*Address all correspondence to: jfeng@lifespan.org

References

[1] Haddy, F. J., Vanhoutte, P. M., & Feletou, M. (2006). Role of potassium in regulating blood flow and blood pressure. American Journal of Physiology-Regulatory, Integrative and Comparative Physiology, *290*(3). https://doi.org/10.1152/ajpregu.00491.2005

[2] Dobson, G. P., Faggian, G., Onorati, F., & Vinten-Johansen, J. (2013). Hyperkalemic cardioplegia for adult and pediatric surgery: end of an era? Frontiers in Physiology, *4*. https://doi.org/10.3389/fphys.2013.00228

[3] Sarkar, M., & Prabhu, V. (2017). Basics of cardiopulmonary bypass. Indian Journal of Anaesthesia, *61*(9), 760. https://doi.org/10.4103/ija.ija_379_17

[4] Maruyama, Y., Chambers, D. J., & Ochi, M. (2013). Future Perspective of Cardioplegic Protection in Cardiac Surgery. Journal of Nippon Medical School, *80*(5), 328-341. https://doi.org/10.1272/jnms.80.328

[5] Weinrich, W. L. (1962). The effects of elective cardiac arrest on ventricular function. Progress in Cardiovascular Diseases, *4*(5), 526-536. https://doi.org/10.1016/s0033-0620(62)80024-3

[6] Mankad, P. S., Chester, A. H., & Yacoub, M. H. (1991). Role of potassium concentration in cardioplegic solutions in mediating endothelial damage. The Annals of Thoracic Surgery, *51*(1), 89-93. https://doi.org/10.1016/0003-4975(91)90457-2

[7] Carvajal C, Goyal A, Tadi P. Cardioplegia. [Updated 2021 Jan 31]. In: StatPearls [Internet]. Treasure Island (FL): StatPearls Publishing; 2021 Jan-. Available from: https://www.ncbi.nlm.nih.gov/books/NBK554463/

[8] Hearse, D. J., Stewart, D. A., & Braimbridge, M. V. (1975). Hypothermic arrest and potassium arrest: metabolic and myocardial protection during elective cardiac arrest. Circulation Research, *36*(4), 481-489. https://doi.org/10.1161/01.res.36.4.481

[9] Ali, J. M., Miles, L. F., Abu-Omar, Y., Galhardo, C., & Falter, F. (2018). Global Cardioplegia Practices: Results from the Global Cardiopulmonary Bypass Survey. The journal of extra-corporeal technology, *50*(2), 83-93.

[10] Mishra, P., Jadhav, R. B., Mohapatra, C. K., Khandekar, J., Raut, C., Ammannaya, G. K., ... Shah, V. (2016). Comparison of del Nido cardioplegia and St. Thomas Hospital solution – two types of cardioplegia in adult cardiac surgery. *Polish Journal of Cardio-Thoracic Surgery*, *4*, 295-299. https://doi.org/10.5114/kitp.2016.64867

[11] Comentale, G., Giordano, R., & Palma, G. (2018). Comparison of the different cardioplegic strategies in cardiac valves surgery: who wins the "arm-wrestling"? Journal of Thoracic Disease, *10*(2), 714-717. https://doi.org/10.21037/jtd.2018.01.133

[12] Sanetra, K., Pawlak, I., & Cisowski, M. (2018). Del Nido cardioplegia – what is the current evidence? Polish Journal of Cardio-Thoracic Surgery, *15*(2), 114-118. https://doi.org/10.5114/kitp.2018.76477

[13] Brown, P. S., Holland, F. W., Parenteau, G. L., & Clark, R. E. (1991). Magnesium ion is beneficial in hypothermic crystalloid cardioplegia. The Annals of Thoracic Surgery, *51*(3), 359-367. https://doi.org/10.1016/0003-4975(91)90845-h

[14] O'Brien, J. D., Howlett, S. E., Burton, H. J., O'Blenes, S. B., Litz, D. S., & Friesen, C. L. (2009). Pediatric Cardioplegia Strategy Results in Enhanced Calcium Metabolism and Lower Serum Troponin T. The Annals of Thoracic Surgery, *87*(5), 1517-1523. https://doi.org/10.1016/j.athoracsur.2009.02.067

[15] Kim, K., Ball, C., Grady, P., & Mick, S. (2014). Use of del Nido Cardioplegia for Adult Cardiac Surgery at the Cleveland Clinic: Perfusion Implications. The journal of extra-corporeal technology, 46(4), 317-323.

[16] Calafiore, A. M., Teodori, G., Mezzetti, A., Bosco, G., Verna, A. M., Di Giammarco, G., & Lapenna, D. (1995). Intermittent antegrade warm blood cardioplegia. The Annals of Thoracic Surgery, *59*(2), 398-402. https://doi.org/10.1016/0003-4975(94)00843-v

[17] Hoyer, A., Lehmann, S., Mende, M., Noack, T., Kiefer, P., Misfeld, M., ... Mohr, F. W. (2017). Custodiol versus cold Calafiore for elective cardiac arrest in isolated aortic valve replacement: a propensity-matched analysis of 7263 patients†. European Journal of Cardio-Thoracic Surgery, *52*(2), 303-309. https://doi.org/10.1093/ejcts/ezx052

[18] Drescher, C., Diestel, A., Wollersheim, S., Berger, F., & Schmitt, K. R. L. (2011). How does hypothermia protect cardiomyocytes during cardioplegic ischemia? European Journal of Cardio-Thoracic Surgery. https://doi.org/10.1016/j.ejcts.2010.12.006

[19] Viana, F. F., Shi, W. Y., Hayward, P. A., Larobina, M. E., Liskaser, F., & Matalanis, G. (2012). Custodiol versus blood cardioplegia in complex cardiac operations: an Australian experience. European Journal of Cardio-Thoracic Surgery, *43*(3), 526-531. https://doi.org/10.1093/ejcts/ezs319

[20] Tissier, R., Ghaleh, B., Cohen, M. V., Downey, J. M., & Berdeaux, A. (2011). Myocardial protection with mild hypothermia. Cardiovascular Research, *94*(2), 217-225. https://doi.org/10.1093/cvr/cvr315

[21] Mauney, M. C., & Kron, I. L. (1995). The physiologic basis of warm cardioplegia. The Annals of Thoracic Surgery, *60*(3), 819-823. https://doi.org/10.1016/0003-4975(95)00422-h

[22] Fan, Y., Zhang, A. M., Xiao, Y. B., Weng, Y. G., & Hetzer, R. (2010). Warm versus cold cardioplegia for heart surgery: a meta-analysis. European journal of cardio-thoracic surgery : official journal of the European Association for Cardio-thoracic Surgery, 37(4), 912-919. https://doi.org/10.1016/j.ejcts.2009.09.030

[23] Fremes SE, Weisel RD, Mickle DA, et al. Myocardial metabolism and ventricular function following cold potassium cardioplegia. J Thorac Cardiovasc Surg. 1985;89(4):531-546.

[24] Brown, W. M., Jay, J. L., Gott, J. P., Huang, A. H., Pan-Chih, Horsley, W. S., ... Guyton, R. A. (1993). Warm blood cardioplegia: Superior protection after acute myocardial ischemia. The Annals of Thoracic Surgery, *55*(1), 32-42. https://doi.org/10.1016/0003-4975(93)90470-3

[25] Ahmed AA, Mahboobi SK. Warm Blood Cardioplegia. [Updated 2021 Jan 29]. In: StatPearls [Internet]. Treasure Island (FL): StatPearls Publishing; 2021 Jan-. Available from: https://www.ncbi.nlm.nih.gov/books/NBK558919/\

[26] Martin, T. D., Craver, J. M., Gott, J. P., Weintraub, W. S., Ramsay, J., Mora, C. T., & Guyton, R. A. (1994). Prospective, randomized trial of retrograde warm blood cardioplegia: Myocardial benefit and neurologic threat. The Annals of

Thoracic Surgery, *57*(2), 298-304. https://doi.org/10.1016/0003-4975(94)90987-3

[27] Rosenkranz ER, Buckberg GD, Laks H, Mulder DG. Warm induction of cardioplegia with glutamate-enriched blood in coronary patients with cardiogenic shock who are dependent on inotropic drugs and intra-aortic balloon support. J Thorac Cardiovasc Surg. 1983;86(4):507-518.

[28] Fan, Y., Zhang, A. M., Xiao, Y. B., Weng, Y. G., & Hetzer, R. (2010). Warm versus cold cardioplegia for heart surgery: a meta-analysis. European journal of cardio-thoracic surgery : official journal of the European Association for Cardio-thoracic Surgery, *37*(4), 912-919. https://doi.org/10.1016/j.ejcts.2009.09.030

[29] Mourad, F. A., Fadala, M. A., Ibrahim, A. A., Ammar, A. M., Elnahas, Y. M., Elghanam, M. A., & Elkilany, I. S. (2016). Myocardial protection during CABG: Warm blood versus cold crystalloid cardioplegia, is there any difference? Journal of the Egyptian Society of Cardio-Thoracic Surgery, *24*(3), 215-222. https://doi.org/10.1016/j.jescts.2016.09.001

[30] Mahrose, R., Shorbagy, M. S., Shahin, K. M., & Elwany, S. E. (2020). Warm blood cardioplegia versus cold crystalloid cardioplegia for coronary artery bypass grafting (CABG) in patients with low ejection fraction. Ain-Shams Journal of Anesthesiology, *12*(1). https://doi.org/10.1186/s42077-020-00069-8

[31] Nardi, P., Pisano, C., Bertoldo, F., Vacirca, S. R., Saitto, G., Costantino, A., ... Ruvolo, G. (2018). Warm blood cardioplegia versus cold crystalloid cardioplegia for myocardial protection during coronary artery bypass grafting surgery. Cell Death Discovery, *4*(1). https://doi.org/10.1038/s41420-018-0031-z

[32] Lopes, J. B., & dos Santos Júnior, C. C. (2017). Coronary Perfusion Pressure during Antegrade Cardioplegia in On-Pump CABG Patients. Brazilian Journal of Cardiovascular Surgery. https://doi.org/10.21470/1678-9741-2017-0035

[33] Habertheuer, A., Kocher, A., Laufer, G., Andreas, M., Szeto, W. Y., Petzelbauer, P., ... Wiedemann, D. (2014). Cardioprotection: A Review of Current Practice in Global Ischemia and Future Translational Perspective. BioMed Research International, *2014*, 1-11. https://doi.org/10.1155/2014/325725

[34] Oliveira, M. A., Brandi, A. C., Santos, C. A., Botelho, P. H., Cortez, J. L., & Braile, D. M. (2014). Modes of induced cardiac arrest: hyperkalemia and hypocalcemia - literature review. Revista Brasileira De Cirurgia Cardiovascular. https://doi.org/10.5935/1678-9741.20140074

[35] Snabaitis, A. K., Shattock, M. J., & Chambers, D. J. (1997). Comparison of Polarized and Depolarized Arrest in the Isolated Rat Heart for Long-term Preservation. Circulation, *96*(9), 3148-3156. https://doi.org/10.1161/01.cir.96.9.3148

[36] Fallouh, H. B., Kentish, J. C., & Chambers, D. J. (2009). Targeting for cardioplegia: arresting agents and their safety. Current Opinion in Pharmacology, *9*(2), 220-226. https://doi.org/10.1016/j.coph.2008.11.012

[37] Rasgado-Flores, H., & Blaustein, M. P. (1987). Na/Ca exchange in barnacle muscle cells has a stoichiometry of 3 Na+/1 Ca2+. American Journal of

Physiology-Cell Physiology, *252*(5). https://doi.org/10.1152/ajpcell.1987.252.5.c499

[38] Cohen, N. M., & Lederer, W. J. (1988). Changes in the calcium current of rat heart ventricular myocytes during development. The Journal of Physiology, *406*(1), 115-146. https://doi.org/10.1113/jphysiol.1988.sp017372

[39] Chambers, D. J., & Hearse, D. J. (1999). Developments in cardioprotection: "polarized" arrest as an alternative to "depolarized" arrest. The Annals of Thoracic Surgery, *68*(5), 1960-1966. https://doi.org/10.1016/s0003-4975(99)01020-6

[40] Tani, M. (1990). Mechanisms of Ca2+ Overload in Reperfused Ischemic Myocardium. Annual Review of Physiology, *52*(1), 543-559. https://doi.org/10.1146/annurev.ph.52.030190.002551

[41] Suleiman, M. S., Halestrap, A. P., & Griffiths, E. J. (2001). Mitochondria: a target for myocardial protection. Pharmacology & Therapeutics, *89*(1), 29-46. https://doi.org/10.1016/s0163-7258(00)00102-9

[42] Cooper, W. A., Duarte, I. G., Thourani, V. H., Nakamura, M., Wang, N.-P., Brown, W. M., ... Guyton, R. A. (2000). Hypothermic circulatory arrest causes multisystem vascular endothelial dysfunction and apoptosis. The Annals of Thoracic Surgery, *69*(3), 696-702. https://doi.org/10.1016/s0003-4975(99)01524-6

[43] Schmitt, J. P., Schröder, J., Schunkert, H., Birnbaum, D. E., & Aebert, H. (2002). Role of apoptosis in myocardial stunning after open heart surgery. The Annals of Thoracic Surgery, *73*(4), 1229-1235. https://doi.org/10.1016/s0003-4975(02)03401-x

[44] Fischer, U., Klass, O., Stock, U., Easo, J., Geissler, H. J., Fischer, J. H., ... Melhorn, U. (2003). Cardioplegic arrest induces apoptosis signal-pathway in myocardial endothelial cells and cardiac myocytes. European Journal of Cardio-Thoracic Surgery, *23*(6), 984-990. https://doi.org/10.1016/s1010-7940(03)00113-1

[45] Anselmi, A., Gaudino, M., Possati, G., Biondi-Zoccai, G. G. L., Nasso, G., Girola, F., & Abbate, A. (2004). Myocardial ischemia, stunning, inflammation, and apoptosis during cardiac surgery: a review of evidence. European Journal of Cardio-Thoracic Surgery, *25*(3), 304-311. https://doi.org/10.1016/j.ejcts.2003.12.003

[46] Misao, J., Hayakawa, Y., Ohno, M., Kato, S., Fujiwara, T., & Fujiwara, H. (1996). Expression of bcl-2 Protein, an Inhibitor of Apoptosis, and Bax, an Accelerator of Apoptosis, in Ventricular Myocytes of Human Hearts With Myocardial Infarction. Circulation, *94*(7), 1506-1512. https://doi.org/10.1161/01.cir.94.7.1506

[47] Tanaka, M., Nakae, S., Terry, R. D., Mokhtari, G. K., Gunawan, F., Balsam, L. B., ... Robbins, R. C. (2004). Cardiomyocyte-specific Bcl-2 overexpression attenuates ischemia-reperfusion injury, immune response during acute rejection, and graft coronary artery disease. Blood, *104*(12), 3789-3796. https://doi.org/10.1182/blood-2004-02-0666

[48] Ramlawi, B., Feng, J., Mieno, S., Szabo, C., Zsengeller, Z., Clements, R., Sodha, N., Boodhwani, M., Bianchi, C., & Sellke, F. W. (2006). Indices of apoptosis activation after blood cardioplegia and cardiopulmonary bypass. Circulation, *114*(1 Suppl), I257–I263. https://doi.org/10.1161/CIRCULATIONAHA.105.000828

[49] Lorenzo, H. K., Susin, S. A., Penninger, J., & Kroemer, G. (1999). Apoptosis inducing factor (AIF): a phylogenetically old, caspase-independent effector of cell death. Cell Death & Differentiation, *6*(6), 516-524. https://doi.org/10.1038/sj.cdd.4400527

[50] Lipton, S. A., & Bossy-Wetzel, E. (2002). Dueling Activities of AIF in Cell Death versus Survival. Cell, *111*(2), 147-150. https://doi.org/10.1016/s0092-8674(02)01046-2

[51] Feng, J., Bianchi, C., Sandmeyer, J. L., Li, J., & Sellke, F. W. (2005). Molecular indices of apoptosis after intermittent blood and crystalloid cardioplegia. Circulation, *112*(9 Suppl), I184–I189. https://doi.org/10.1161/CIRCULATIONAHA.104.526160

[52] Ruel, M., Khan, T. A., Voisine, P., Bianchi, C., & Sellke, F. W. (2004). Vasomotor dysfunction after cardiac surgery. European Journal of Cardio-Thoracic Surgery, *26*(5), 1002-1014. https://doi.org/10.1016/j.ejcts.2004.07.040

[53] Quignard, J.-F., Félétou, M., Thollon, C., Vilaine, J.-P., Duhault, J., & Vanhoutte, P. M. (1999). Potassium ions and endothelium-derived hyperpolarizing factor in guinea-pig carotid and porcine coronary arteries. British Journal of Pharmacology, *127*(1), 27-34. https://doi.org/10.1038/sj.bjp.0702493

[54] Brayden, J. E. (1996). POTASSIUM CHANNELS IN VASCULAR SMOOTH MUSCLE. Clinical and Experimental Pharmacology and Physiology, *23*(12), 1069-1076. https://doi.org/10.1111/j.1440-1681.1996.tb01172.x

[55] Dick, G. M., & Tune, J. D. (2010). Role of potassium channels in coronary vasodilation. Experimental Biology and Medicine, *235*(1), 10-22. https://doi.org/10.1258/ebm.2009.009201

[56] He, G.-W., & Yang, C.-Q. (1996). Hyperkalemia alters endothelium-dependent relaxation through non—nitric oxide and noncyclooxygenase pathway: A mechanism for coronary dysfunction due to cardioplegia. The Annals of Thoracic Surgery, *61*(5), 1394-1399. https://doi.org/10.1016/0003-4975(96)00086-0

[57] Wu, M., Dong, Y.-Y., Yang, Q., Yim, A. P. C., & He, G.-W. (2005). Cellular Electrophysiological and Mechanical Effects of Celsior Solution on Endothelial Function in Resistance Coronary Arteries. Transplantation, *80*(12), 1765-1772. https://doi.org/10.1097/01.tp.0000183961.17370.71

[58] Furchgott, R. F., & Zawadzki, J. V. (1980). The obligatory role of endothelial cells in the relaxation of arterial smooth muscle by acetylcholine. Nature, *288*(5789), 373-376. https://doi.org/10.1038/288373a0

[59] Tyers G. F. (1997). Microvascular reactivity after cardioplegia. The Annals of thoracic surgery, *63*(2), 598-599. https://doi.org/10.1016/s0003-4975(97)83391-7

[60] Parolari, A., Rubini, P., Cannata, A., Bonati, L., Alamanni, F., Tremoli, E., & Biglioli, P. (2002). Endothelial damage during myocardial preservation and storage. The Annals of Thoracic Surgery, *73*(2), 682-690. https://doi.org/10.1016/s0003-4975(01)03029-6

[61] Feng, J., Liu, Y., Khabbaz, K. R., Hagberg, R., Sodha, N. R., Osipov, R. M., & Sellke, F. W. (2010). Endothelin-1-induced contractile responses of human coronary arterioles via endothelin-A receptors and PKC-α signaling pathways. Surgery, *147*(6), 798-804. https://doi.org/10.1016/j.surg.2009.11.016

[62] Feng, J., Liu, Y., Chu, L. M., Clements, R. T., Khabbaz, K. R., Robich, M. P., ... Sellke, F. W. (2011). Thromboxane-Induced Contractile Response of Human Coronary Arterioles Is Diminished After Cardioplegic Arrest. The Annals of Thoracic Surgery, *92*(3), 829-836. https://doi.org/10.1016/j.athoracsur.2011.04.049

[63] Sellke FW, Shafique T, Ely DL, Weintraub RM. Coronary endothelial injury after cardiopulmonary bypass and ischemic cardioplegia is mediated by oxygen-derived free radicals. Circulation. 1993 Nov;88(5 Pt 2):II395-II400. PMID: 8222185.

[64] Sellke, F. W., Friedman, M., Dai, H. B., Shafique, T., Schoen, F. J., Weintraub, R. M., & Johnson, R. G. (1993). Mechanisms causing coronary microvascular dysfunction following crystalloid cardioplegia and reperfusion. Cardiovascular Research, *27*(11), 1925-1932. https://doi.org/10.1093/cvr/27.11.1925

[65] Chenoweth, D. E., Cooper, S. W., Hugli, T. E., Stewart, R. W., Blackstone, E. H., & Kirklin, J. W. (1981). Complement Activation during Cardiopulmonary Bypass. New England Journal of Medicine, *304*(9), 497-503. https://doi.org/10.1056/nejm198102263040901

[66] Feng, J., & Sellke, F. (2016). Microvascular dysfunction in patients with diabetes after cardioplegic arrest and cardiopulmonary bypass. Current Opinion in Cardiology, *31*(6), 618-624. https://doi.org/10.1097/hco.0000000000000340

[67] Giacinto, O., Satriano, U., Nenna, A., Spadaccio, C., Lusini, M., Mastroianni, C., ... Chello, M. (2019). Inflammatory Response and Endothelial Dysfunction Following Cardiopulmonary Bypass: Pathophysiology and Pharmacological Targets. Recent Patents on Inflammation & Allergy Drug Discovery, *13*(2), 158-173. https://doi.org/10.2174/1872213x13666190724112644

[68] Jellinek, M., Standeven, J. W., Menz, L. J., Hahn, J. W., & Barner, H. B. (1981). Cold blood potassium cardioplegia: Effects of increasing concentrations of potassium. The Journal of Thoracic and Cardiovascular Surgery, *82*(1), 26-37. https://doi.org/10.1016/s0022-5223(19)39383-3

[69] Hoover, E. L., Pett, S. B., Eldor, A., Alonso, D., Subramanian, V. A., Weksler, B., et al. (1981). The effects of crystalloid potassium cardioplegic solution on arterialized canine vein grafts. Assessment of chronic prostacyclin production and histopathologic alterations. Circulation 64(2 Pt 2), II96–II100.

[70] Olinger, G. N., Boerboom, L. E., Bonchek, L. I., Hutchinson, L. D., & Kissebah, A. H. (1983). Hyperkalemia in cardioplegic solutions causing increased cholesterol accumulation in vein grafts. The Journal of Thoracic and Cardiovascular Surgery, *85*(4), 590-594. https://doi.org/10.1016/s0022-5223(19)37544-0

[71] Bianchi, C., Araujo, E. G., Sato, K., & Sellke, F. W. (2001). Biochemical and Structural Evidence for Pig Myocardium Adherens Junction Disruption by Cardiopulmonary Bypass. Circulation, *104*(suppl 1). https://doi.org/10.1161/hc37t1.094519

[72] Feng, J., Liu, Y., Sabe, A. A., Sadek, A. A., Singh, A. K., Sodha, N. R., & Sellke, F. W. (2015). Differential impairment of adherens-junction expression/phosphorylation after cardioplegia in diabetic versus non-diabetic patients. European Journal of

Cardio-Thoracic Surgery, *49*(3), 937-943. https://doi.org/10.1093/ejcts/ezv202

[73] Li, J.-M., & Shah, A. M. (2004). Endothelial cell superoxide generation: regulation and relevance for cardiovascular pathophysiology. American Journal of Physiology-Regulatory, Integrative and Comparative Physiology, *287*(5). https://doi.org/10.1152/ajpregu.00124.2004

[74] Breemen, C., Skarsgard, P., Laher, I., McManus, B., & Wang, X. (1997). ENDOTHELIUM-SMOOTH MUSCLE INTERACTIONS IN BLOOD VESSELS. Clinical and Experimental Pharmacology and Physiology, *24*(12), 989-992. https://doi.org/10.1111/j.1440-1681.1997.tb02737.x

[75] Sohn, H.-Y., Keller, M., Gloe, T., Morawietz, H., Rueckschloss, U., & Pohl, U. (2000). The Small G-protein Rac Mediates Depolarization-induced Superoxide Formation in Human Endothelial Cells. Journal of Biological Chemistry, *275*(25), 18745-18750. https://doi.org/10.1074/jbc.m000026200

[76] Ishikawa, Y., & Sasakawa, S. (1987). Membrane potential of stored platelets and its effect on platelet functions. Thrombosis Research, *45*(3), 265-273. https://doi.org/10.1016/0049-3848(87)90194-0

[77] Krötz Florian, Sohn, H. Y., Keller, M., Gloe, T., Bolz, S. S., Becker, B. F., & Pohl, U. (2002). Depolarization of Endothelial Cells Enhances Platelet Aggregation Through Oxidative Inactivation of Endothelial NTPDase. Arteriosclerosis, Thrombosis, and Vascular Biology, *22*(12), 2003-2009. https://doi.org/10.1161/01.atv.0000043454.08172.51

[78] Soori, R., Dixit, A., & Tewari, P. (2018). Refractory hypokalemia while weaning off bypass. Annals of Cardiac Anaesthesia, *21*(3), 311. https://doi.org/10.4103/aca.aca_196_17

[79] Vasko, K. A., DeWall, R. A., & Riley, A. M. (1973). Hypokalemia: Physiological Abnormalities During Cardiopulmonary Bypass. The Annals of Thoracic Surgery, *15*(4), 347-353. https://doi.org/10.1016/s0003-4975(10)65313-1

[80] Su, H. (1986). Hypokalemia in Cardiac Surgery under Cardiopulmonary Bypass (CPB). *International Practice in Cardiothoracic Surgery*, 145-160. https://doi.org/10.1007/978-94-009-4259-2_16

[81] Polderman, K. H., & Girbes, A. R. J. (2004). Severe electrolyte disorders following cardiac surgery: a prospective controlled observational study. Critical Care, *8*(6). https://doi.org/10.1186/cc2973

[82] Peretto, G., Durante, A., Limite, L. R., & Cianflone, D. (2014). Postoperative Arrhythmias after Cardiac Surgery: Incidence, Risk Factors, and Therapeutic Management. Cardiology Research and Practice, *2014*, 1-15. https://doi.org/10.1155/2014/615987

[83] Al-Ghamdi, B. (2017). Arrhythmias Post Coronary Artery Bypass Surgery. Coronary Artery Bypass Graft Surgery. https://doi.org/10.5772/intechopen.70423

[84] Reves, J. G., Karp, R. B., Buttner, E. E., Tosone, S., Smith, L. R., Samuelson, P. N., Kreusch, G. R., & Oparil, S. (1982). Neuronal and adrenomedullary catecholamine release in response to cardiopulmonary bypass in man. Circulation, *66*(1), 49-55. https://doi.org/10.1161/01.cir.66.1.49

[85] Tan, C.-K., Glisson, S. N., El-Etr, A. A., & Ramakrishnaiah, K. B. (1976). Levels of circulating norepinephrine and epinephrine before, during, and after

cardiopulmonary bypass in man. The Journal of Thoracic and Cardiovascular Surgery, *71*(6), 928-931. https://doi.org/10.1016/s0022-5223(19)40133-5

[86] Moratinos, J., & Reverte, M. (1993). Effects of catecholamines on plasma potassium: the role of alpha- and beta-adrenoceptors. Fundamental & Clinical Pharmacology, *7*(3-4), 143-153. https://doi.org/10.1111/j.1472-8206.1993.tb00228.x

[87] Li, T., & Vijayan, A. (2014). Insulin for the treatment of hyperkalemia: a double-edged sword? Clinical Kidney Journal, *7*(3), 239-241. https://doi.org/10.1093/ckj/sfu049

[88] Buse, S., Blancher, M., Viglino, D., Pasquier, M., Maignan, M., Bouzat, P., Annecke, T., & Debaty, G. (2017). The impact of hypothermia on serum potassium concentration: A systematic review. Resuscitation, *118*, 35-42. https://doi.org/10.1016/j.resuscitation.2017.07.003

[89] Brauer, S. D., Applegate, R. L., Jameson, J. J., Hay, K. L., Lauer, R. E., Herrmann, P. C., & Bull, B. S. (2013). Association of Plasma Dilution With Cardiopulmonary Bypass-Associated Bleeding and Morbidity. Journal of Cardiothoracic and Vascular Anesthesia, *27*(5), 845-852. https://doi.org/10.1053/j.jvca.2013.01.011

[90] Knudson, J. D., Lowry, A. W., Price, J. F., & Moffett, B. S. (2012). Response to Intravenous Potassium Chloride Supplementation in Pediatric Cardiac Intensive Care Patients. Pediatric Cardiology, *34*(4), 887-892. https://doi.org/10.1007/s00246-012-0565-4

[91] Khoo, M. S. C., Braden, G. L., Deaton, D., Owen, S., Germain, M., O'Shea, M., Mulhern, J., Rousou, J., Flack, J., & Engleman, R. (2003). Outcome and complications of intraoperative hemodialysis during cardiopulmonary bypass with potassium-rich cardioplegia. American Journal of Kidney Diseases, *41*(6), 1247-1256. https://doi.org/10.1016/s0272-6386(03)00369-x

[92] Lema, G. F., Tesema, H. G., Fentie, D. Y., & Arefayne, N. R. (2019). Evidence-based perioperative management of patients with high serum potassium level in resource-limited areas: A systematic review. International Journal of Surgery Open, *21*, 21-29. https://doi.org/10.1016/j.ijso.2019.10.006

[93] Putcha, N., & Allon, M. (2007). Management of Hyperkalemia in Dialysis Patients. Seminars in Dialysis, *20*(5), 431-439. https://doi.org/10.1111/j.1525-139x.2007.00312.x

[94] Cohen, N. M., Wise, R. M., Wechsler, A. S., & Damiano, R. J. (1993). Elective cardiac arrest with a hyperpolarizing adenosine triphosphate–sensitive potassium channel opener. The Journal of Thoracic and Cardiovascular Surgery, *106*(2), 317-328. https://doi.org/10.1016/s0022-5223(19)34131-5

[95] Cohen, N. M., Damiano, R. J., & Wechsler, A. S. (1995). Is there an alternative to potassium arrest? The Annals of Thoracic Surgery, *60*(3), 858-863. https://doi.org/10.1016/0003-4975(95)00210-c

[96] Chambers, D. J. (1999). Polarization and myocardial protection. Current Opinion in Cardiology, *14*(6), 495. https://doi.org/10.1097/00001573-199911000-00008

Chapter 5

Potassium in Solid Cancers

Jessica Iorio, Lisa Lastraioli and Elena Lastraioli

Abstract

Electrolyte disorders are a frequent finding in cancer patients. In the majority of cases the etiologies of such disorders are common to all cancer types (i.e. diuretic-induced hyponatremia or hypokalemia). Sometimes, electrolyte disorders are caused by paraneoplastic syndromes or are due to cancer therapy. Potassium is one of the most important electrolytes of the human body since it is involved in the regulation of muscle contraction, maintenance of the integrity of the skeleton, blood pressure and nerve transmission as well as in the normal function of cells. Potassium homeostasis is strictly regulated since the gap between the recommended daily dietary intake (120 mEq/day) and the levels stored in the extracellular fluid (around 70 mEq) is huge. Alterations of potassium homeostasis are frequent in cancer patients as well alterations in potassium channels, the transmembrane proteins that mediate potassium fluxes within the cells. The present chapter is focused on the clinical significance of potassium homeostasis and potassium channels in patients with solid tumors.

Keywords: potassium, electrolyte homeostasis, solid cancers, potassium channels, channelopathies

1. Introduction

Potassium has a great importance within the cell as it is one of the main determinants of membrane potential in nervous and muscle cells (both skeletal and cardiac). Therefore, it is not surprising that in humans potassium blood levels are strictly controlled and even slight deviations from the physiological range can cause severe clinical conditions.

The amount of potassium in the human body is 3000–4000 mEq (corresponding to 50–55 mEq/Kg). Potassium levels in the extracellular fluid are quite low (roughly 70 mEq corresponding to 2% of the amount absorbed through the intestine) while the majority of K^+ (98%) is stored in the body tissues under a tight hormonal control. This great gap between extracellular fluid and tissues is responsible of the restricted range of K^+ levels in the plasma (3.5–5.5 mEq/L) that is maintained through the following mechanisms:

- Activity of Na-K ATPase. The Na-K pump actively extrudes three Na^+ from the cells exchanging them with two K^+ that enter the cells against gradient [1]. Such mechanism makes it possible the preservation of the physiological electrolyte concentration as well as the maintenance of the normal volume of extracellular fluid. It is also essential for the solute passage through renal and intestinal epithelia.

- Elimination of potassium dietary intake through feces and urine.

A scheme of potassium intake, absorption and elimination is reported in **Figure 1**.

Potassium intake from diet is mainly due to the consumption of K^+-rich food such as bananas, avocados, grapes and salmon, among others. Once introduced in the body as food, potassium is rapidly conveyed towards intestine, where it is absorbed for its majority (90%). It has been suggested that absorption in the intestine is a passive process, guided by electrochemical gradients differently from the one taking place in the kidney [2].

Insulin is one of the main regulators of potassium homeostasis, modulating the exchange between extracellular fluid and tissues such as muscle, through the activation of the Na- K ATPase [3]. Moreover, in muscular tissue during exercise, a rise in potassium levels causes the stimulation of Na-K ATPase thus enhancing K^+ uptake [4], although the primary upregulation of the ATPase is due to adrenergic stimulation.

From the extracellular fluid, K^+ is also conveyed towards kidney where in physiological conditions it is partly reabsorbed and for the major portion eliminated with urine (90%) in the distal renal tubule, directed by mineralocorticoids and Na-K ATPase. The mechanisms of K^+ renal reabsorption are outside the scope of this chapter and can be deepened in several reviews (see for example [5]).

K^+ distribution across plasma-membranes is regulated by pancreatic hormones, acid–base homeostasis alterations and by the autonomous nervous system.

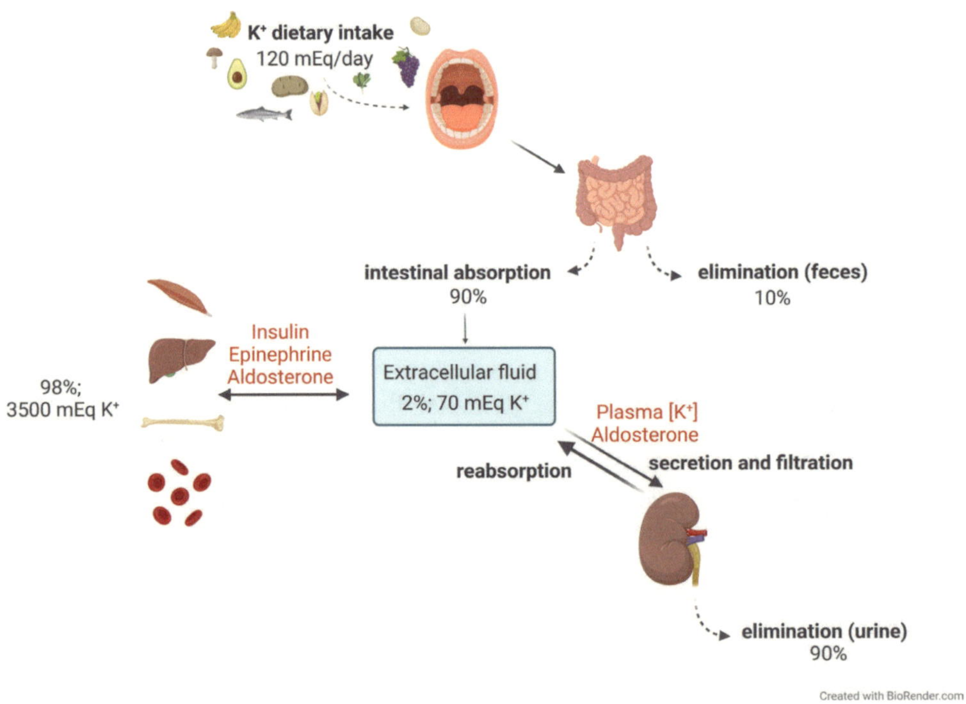

Figure 1.
Schematic diagram of K^+ intake, fluxes, storage and elimination.

Moreover, potassium itself regulates somehow its own excretion and redistribution through the stimulation of insulin and aldosterone secretion as well as the increase of Na-K ATPase activity in the distal nephron. On the whole, K^+ homeostasis is due to the interaction of two different systems, one based on K^+ cellular uptake and the other based mainly on renal excretion and, at a minor extent, on gastrointestinal elimination. These two systems have an important difference since the redistribution of potassium between intra- and extra-cellular compartments occurs rapidly (it takes minutes) while renal response to K^+ variations in plasma is quite slow (it takes hours to be activated and days to be completed). Moreover, since the biological effects depend on the ratio between the external and the internal K^+ concentration, the regulation of the internal distribution must be extremely efficient. A tight association exists between pH and kaliemia, therefore acidosis causes an increase of potassium plasma level while alkalosis induces hypokalemia. Another pathological condition that might cause hyperkalemia is the increase of osmolarity (due to hyperglycemia and hypernatremia).

2. Electrolyte disorders in human cancer

Cancer patients are characterized by complex alterations encompassing the whole organism. Among them electrolyte disorders represent quite frequent complications in such patients [6, 7] and are mainly due to alterations of sodium, potassium, calcium and magnesium plasma levels. In the majority of the patients these unbalances are asymptomatic but in some subjects they can contribute to worsen the clinical conditions and therefore must be treated. The electrolyte derangements adversely affect survival [8] and can disrupt cancer treatment [9–11]. There are many potential causes of electrolyte derangements in cancer patients that might be related either to a particular cancer or to any specific therapy. Actually, the picture might be even more complicated since these electrolyte disorders can be primary or secondary and active treatment can improve prognosis.

3. Potassium homeostasis alterations in human cancer

In cancer patients both a reduction (hypokalemia) or an increase (hyperkalemia) in potassium levels might be observed.

Hypokalemia is a condition of low K^+ plasma levels (< 3.5 mEq/L) that might have different causes in cancer patients. The principal causes of hypokalemia are reported in **Table 1**.

In cancer patients, the main cause of hypokalemia is a reduced K^+ intake due to malnutrition and malabsorption. In particular cases (i.e. Neuroendocrine Tumors), hypokalemia could be provoked by secretive diarrhea that causes potassium losses [12, 13]. Other tumors cause hypokalemia with the ectopic secretion of hormones (cortisol, ACTH and mineralocorticoids) [14].

In other cases, hypokalemia could be a secondary effect of cancer therapy, since some agents might cause diarrhea or vomiting thus leading to K^+ loss. Different cancer therapy agents (such as cisplatin, anti-EGFR agents and mTOR inhibitors, for instance) could induce renal damage associated to hypokalemia due to K^+ loss caused by tubular toxicity. For this reason, before starting therapy, renal function should be evaluated in order to prevent additional renal damage [15].

Decreased K⁺ intake	Starvation		
	Clay ingestion		
Redistribution into cells	Acid-Base		
		Metabolic alkalosis	
	Hormonal		
		Insulin	
		β2 adrenergic agonists	
		α adrenergic antagonists	
	Anabolic state	Vitamin B12 or folic acid	
		Granulocyte-macrophage colony stimulating factor	
		Total parenteral nutrition	
	Other	Pseudo-hypokalemia	
		Hypothermia	
		Hypokalemic periodic paralysis	
		Barium toxicity	
Increased loss	**Nonrenal**	**Gastrointestinal loss (diarrhea)**	
		Integumentary loss (sweat)	
	Renal	**Increased distal flow (diuretics....)**	
		Increased K+ secretion	**Mineralocorticoid excess**
			Distal delivery of non-reabsorbed anions (vomiting, nasogastric suction....)
	Other	Amphotericin B	
		Liddle' s syndrome	
		Hypomagnesemia	

The causes primarily represented in cancer patients are indicated in bold.

Table 1.
Main causes of hypokalemia.

Also, the simultaneous treatment with different agents (for example thiazide diuretics and glucocorticoids) might lead to hypokalemia because they promote potassium renal losses. Finally, the concomitant presence of endocrine dysfunctions might cause hypokalemia due to glucocorticoids or mineralocorticoids excess [14, 16].

Hyperkalemia is a condition of increased K^+ plasma levels (> 5.5 mEq/L). The principal causes of hyperkalemia are reported in **Table 2**.

The main causes of hyperkalemia in cancer patients are represented by reduced renal elimination caused by renal failure and redistribution within extracellular compartment.

In cancer patients there are different conditions than can induce hyperkalemia [1]. For instance, highly proliferative tumors can develop lysis syndrome after anticancer therapy, thus causing hyperkalemia. As in the case of hypokalemia, also treatment

Renal failure		
Decreased distal flow		
Decreased K^+ secretion		
	Impaired Na^+ reabsorption	
		Primary hypoaldosteronism
		Secondary hypoaldosteronism
		Resistance to aldosterone
Enhanced Cl^- reabsorption **NCC channel at the distal tubule**	Gordon's syndrome	
	Cyclosporine	

The causes primarily represented in cancer patients are indicated in bold.

Table 2.
Main causes of hyperkalemia.

with chemotherapeutic agents such as platinum compounds might lead to hyperkalemia due to renal injury. Patients treated with a combination of drugs (i.e. potassium sparing diuretics, NSAIDs, angiotensin-converting enzyme inhibitors) might develop hyperkalemia. Also, the occurrence of concomitant disease (i.e. diabetes mellitus, sepsis, renal failure) or the need of parenteral nutrition might cause hyperkalemia [17, 18]. An important cause of hyperkalemia is represented by Tumor Lysis Syndrome, a serious condition characterized by hyperkalemia, hyperphosphatemia, hypocalcemia, hyperuricemia and hyperazotemia (**Figure 2**).

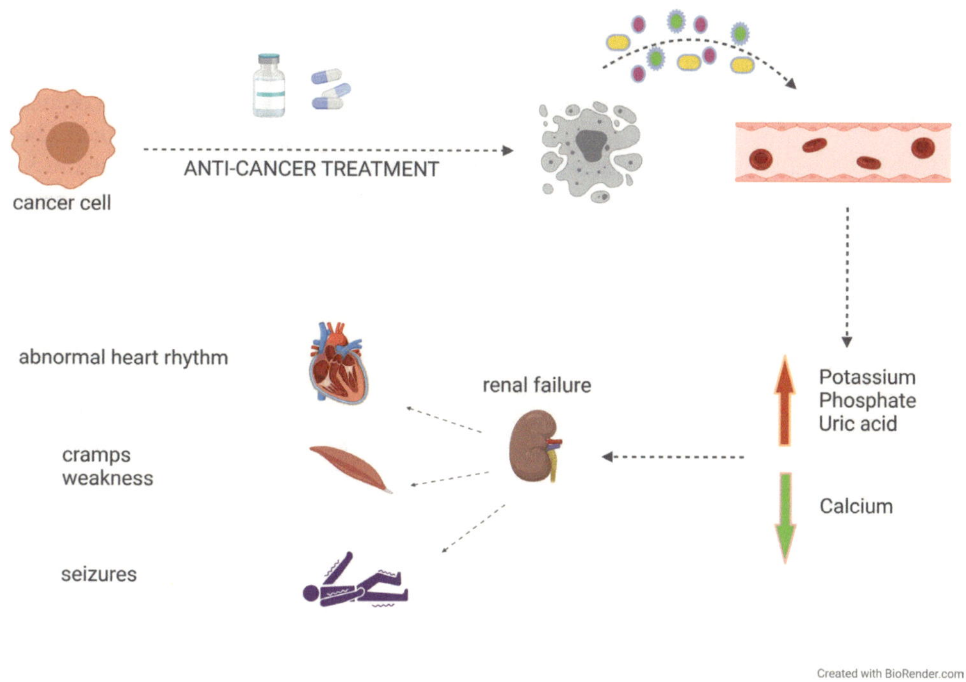

Figure 2.
Schematic diagram of tumor lysis syndrome.

TLS generally occurs within 72 hours from the therapy beginning and it is due to necrosis and subsequent release of cell content into the bloodstream [19]. In particular, there is an increase of potassium, phosphate and uric acid plasma levels while calcium levels are reduced. As a consequence, calcium phosphate precipitates in renal tubules, thus causing kidney injury and failure. Renal injury causes a worsening of metabolic disorders and hyperkalemia that can lead to arrhythmias, seizures, weakness and muscle cramps. Some cancer patients have a higher risk of developing TLS: for example, patients with a large tumor burden, rapidly growing tumors and tumors highly responsive to therapy (either standard chemotherapy, hormonal, radiation and targeted therapy). All these factors should be carefully evaluated before starting the treatment [20]. TLS generally occurs after a week of treatment but it can also develop spontaneously, for example in childhood cancers [21]. Moreover, patients suffering from additional renal dysfunctions have a higher risk of developing TLS.

Other causes of hyperkalemia in cancer patients are represented by adrenal insufficiency (Addison's disease) due to metastases in the adrenal gland especially in patients with advanced tumor of the breast and lung. Nevertheless, although the frequency of adrenal metastases is high, adrenal failure is rare due to the use of corticosteroids [22].

3.1 Potassium in paraneoplastic syndromes

According to the National Cancer Institute (NCI, Dictionary of Cancer Terms, https://www.cancer.gov/publications/dictionaries/cancer-terms) a paraneoplastic syndrome is defined as follows: "A group of symptoms that may develop when substances released by some cancer cells disrupt the normal function of surrounding cells and tissue". This definition implies that a cancer patient may develop symptoms that are not directly attributable to tumor invasion or compression [23]. Paraneoplastic syndromes may have impact on different organs, such as endocrine, neurologic, hematologic, rheumatologic and dermatologic systems. Such group of disorders may be due to the ectopical secretion of hormones and peptides (endocrine paraneoplastic syndromes) or immune cross-reactivity (neurologic paraneoplastic syndromes) [24].

The only endocrine paraneoplastic syndrome involving K^+ alteration is Cushing syndrome, characterized by hypercortisolism [25]. The syndrome is caused by the secretion of adrenocorticotropic hormone or corticotropin-releasing factor from tumor cells [26, 27] followed by cortisol secretion from the adrenal glands. The main features of Cushing syndrome are reported in **Table 3**.

Clinical features	Laboratory findings	Associated tumors	References
Muscle weakness; Peripheral edema; Hypertension; Weight gain; Centripetal fat distribution	Hypokalemia (<3.0 mEq); High cortisol baseline level (>29 μg/dL); Urinary free cortisol level > 47 μg/24 h Midnight ACTH level > 100 ng/L	Neuroendocrine lung tumors; Small cell lung cancer; Thymoma; Medullary thyroid cancer; Gastrointestinal tract cancer; Ovarian cancer; Pancreatic cancer; Adrenal cancer	[26–29]

Table 3.
Main features of Cushing syndrome.

In contrast to other paraneoplastic syndromes (i.e. syndrome of inappropriate antidiuretic hormone secretion and hypercalcemia), patients often develop Cushing syndrome's symptoms before cancer is diagnosed. Moreover, the reappearance of Cushing syndrome may precede tumor recurrence [24].

4. Potassium channels

The multi-gene family of potassium channels is made of several subfamilies (**Figure 3**): calcium-activated channels, voltage-gated channels, inward rectifiers and two-pore domains (reviewed in [30, 31]).

Calcium-activated potassium channels are represented by two classes of proteins: "small- and intermediate- conductance" (SK) and "high-conductance" (BK) potassium channels. SK channels assemble as tetramers and each monomer is composed by six transmembrane domains (S1-S6) with a central pore in the S5-S6 region. Similarly, BK channels are organized as tetramers with α and β subunits; in this case the pore is formed by α subunits.

Voltage-gated potassium channels are composed by four subunits (each constituted by six transmembrane domains termed S1-S6) surrounding an aqueous pore. In these channels, the pore (P) is formed by a loop between S5 and S6. S4 domain acts as voltage sensor.

Inward rectifier potassium channels are tetramers, with each monomer characterized by only two transmembrane domains linked by a P-loop.

The last subfamily is represented by Two-pore domains channels, made of four transmembrane domains with two regions forming the aqueous pores.

4.1 Potassium channels in human cancer

Among ion channels, those selective for K^+ channels are the most frequently deregulated in cancers. Several reports have been published over the years, highlighting

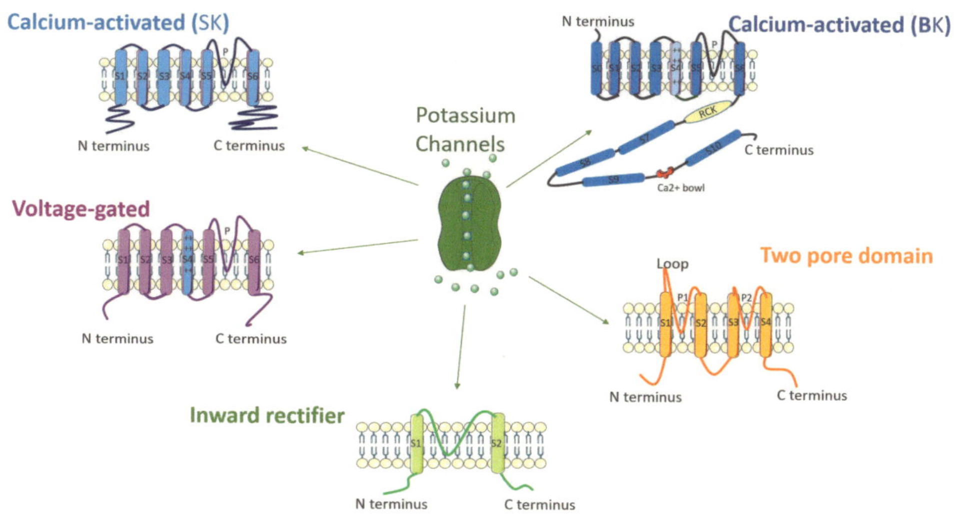

Figure 3.
Potassium channels subfamilies.

K^+ channels' relevance in human solid cancers [31]. For example, in colorectal cancer the presence of the transcripts of *KCNA3, KCNA5, KCNC1, KCNH1* [32–34], *KCNH2* [35] and *KCNK9* [36] was reported.

KCa1.1 channels (also indicated as BK channels) are encoded by the *KCNMA1* gene, located in 10q22, and belong to the high conductance subgroup of the calcium-activated subfamily. For these channels a clinical relevance in breast and prostate cancer has been demonstrated [37–41]. In breast cancer, *KCNMA1* gene amplification occurs in invasive ductal tumors, and it is associated with high grade, high stage and unfavorable prognosis [37]. Moreover, KCa1.1 expression positively correlates with estrogen receptors [41] and the channel is expressed at higher levels in breast cancers developing metastases to brain [38]. Similarly, in prostate cancer *KCNMA1* is more frequently amplified in late-stage tumors [39] and represents a promising diagnostic biomarker since it is over-expressed in tumors with Gleason score equal to 5–6 [40].

KCa3.1, encoded by the *KCNN4* gene, is another member of the calcium-activated subfamily that is frequently upregulated in high grade breast cancers [42], small cell lung cancer [43], colorectal cancer [44] and pancreatic ductal adenocarcinomas [45]. In small cell lung cancer, *KCNN4* hypomethylation was shown to be a negative prognostic factor [43].

Kv1.3 (also named *KCNA3*) is a channel belonging to the voltage gated subfamily that was shown to be overexpressed in Gleason score 5–6 prostate cancer samples with respect to Gleason Score 8–9 patients [46], being down-regulated in high grade tumors [47]. In pancreatic cancer Kv1.3 is down-expressed with respect to healthy pancreas and such downregulation could be due to the methylation of the promoter of the gene and it was also associated with the development of metastases [48].

Kv7.1 (also named KCNQ1) was shown to be over-expressed in lung tumors and to regulate cell proliferation and migration [49].

Kv10.1 (KCNH1) was shown to be expressed in esophageal squamous cell carcinoma and it was proposed as an independent negative prognostic factor [50]. Kv10.1 is highly expressed also in colorectal cancers where it is associated with lymph node metastasis, tumor size, Dukes staging and was therefore proposed as a prognostic marker [51]. Also in gastric cancer it was demonstrated that high Kv10.1 expression is associated with lymph node metastasis and higher stage [52].

Kv11.1 (KCNH2) was found to be expressed in a high percentage of esophageal squamous cell carcinoma samples with respect to healthy esophageal squamous epithelium [53]. Similarly, in esophageal adenocarcinomas it was shown that Kv11.1 as well as the corresponding gene are expressed [54] and more recently it was shown that preneoplastic lesions expressing high levels of Kv11.1 have a significantly higher risk to progress towards adenocarcinoma [55]. Kv11.1 is also associated with TNM stage, grading, serosal and venous invasion, Lauren intestinal type and VEGF-A expression in gastric cancer [56, 57]. In colorectal cancers, the overexpression of Kv11.1 regulates cell invasion [35]. Moreover, in TNM I and II samples Kv11.1 presence associated with Glut-1 absence represents an independent negative prognostic factor [58]. More recently, it was shown that the concomitant expression of hERG1 and HIF-2α represents a prognostic biomarker and it can be used to select metastatic CRC patients suitable to be treated with Bevacizumab [59]. In pancreatic cancer Kv11.1 is associated to aggressive behavior and poorer prognosis [60].

Kir3.1 (KCNJ3) channels belong to the inward rectifier subfamily and it was shown that they positively correlated with lymph node metastases in breast cancer [61].

Several members of the Two Pore Domain family have been shown to be overexpressed in human tumors [62]. Quite recently, **K2P6.1** (KCNK6) was shown to be expressed in breast cancer and to promote proliferation, migration and invasion [63]. The gene encoding for **K2P9.1** (*KCNK9,* located on 8q23.4) was proven to be amplified in breast cancer [64]. Another member of the same family, **K2P5.1** (KCNK5) is induced by estrogens in ER-positive breast cancer cells [65]. *K2p 2.1* is overexpressed in prostate cancer, where it regulates cell proliferation [66].

5. Conclusion

Potassium is one of the most important mineral for humans, therefore its homeostasis is under tight control. Alterations of potassium homeostasis are frequent in cancer patients as well as alterations in potassium channels, and the clinical management of such patients must take into account the possibility for the patient of developing hypo- and hyperkaliemia to define a proper protocol of treatment.

Conflict of interest

The authors declare no conflict of interest.

Author details

Jessica Iorio[1], Lisa Lastraioli[2] and Elena Lastraioli[1*]

1 Department of Experimental and Clinical Medicine, University of Florence, Florence, Italy

2 Intermediate Care Unit, AOU Careggi, Florence, Italy

*Address all correspondence to: elena.lastraioli@unifi.it

References

[1] Clausen, T.; Everts, M. Regulation of the Na,K-pump in skeletal muscle. Kidney Int. **1989**, *35*, 1-13, doi:10.1038/KI.1989.1.

[2] Agarwal, R.; Afzalpurkar, R.; Fordtran, J. Pathophysiology of potassium absorption and secretion by the human intestine. Gastroenterology **1994**, *107*, 548-571, doi:10.1016/0016-5085(94)90184-8.

[3] McKenna, M.; Gissel, H.; Clausen, T. Effects of Electrical Stimulation and Insulin on Na+–K+-ATPase ([3H] Ouabain Binding) in Rat Skeletal Muscle. J. Physiol. **2003**, *547*, 567, doi:10.1113/JPHYSIOL.2003.034512.

[4] DiFranco, M.; Hakimjavadi, H.; Lingrel, J.; Heiny, J. Na,K-ATPase α2 activity in mammalian skeletal muscle T-tubules is acutely stimulated by extracellular K+. J. Gen. Physiol. **2015**, *146*, 281-294, doi:10.1085/JGP.201511407.

[5] McDonough, A.; Youn, J. Potassium Homeostasis: The Knowns, the Unknowns, and the Health Benefits. Physiology (Bethesda). **2017**, *32*, 100-111, doi:10.1152/PHYSIOL.00022.2016.

[6] Berardi, R.; Torniai, M.; Lenci, E.; Pecci, F.; Morgese, F.; Rinaldi, S. Electrolyte disorders in cancer patients: a systematic review. *J. Cancer Metastasis Treat.* **2019**, *5*, doi:10.20517/2394-4722.2019.008.

[7] Rosner, M.; Dalkin, A. Electrolyte disorders associated with cancer. Adv. Chronic Kidney Dis. **2014**, *21*, 7-17, doi:10.1053/J.ACKD.2013.05.005.

[8] Ingles Garces, A.; Ang, J.; Ameratunga, M.; Chénard-Poirier, M.; Dolling, D.; Diamantis, N.; Seeramreddi, S.; Sundar, R.; de Bono, J.; Lopez, J.; et al. A study of 1088 consecutive cases of electrolyte abnormalities in oncology phase I trials. Eur. J. Cancer **2018**, *104*, 32-38, doi:10.1016/J.EJCA.2018.08.019.

[9] Berardi, R.; Rinaldi, S.; Caramanti, M.; Grohè, C.; Santoni, M.; Morgese, F.; Torniai, M.; Savini, A.; Fiordoliva, I.; Cascinu, S. Hyponatremia in cancer patients: Time for a new approach. Crit. Rev. Oncol. Hematol. **2016**, *102*, 15-25, doi:10.1016/J.CRITREVONC.2016.03.010.

[10] Liamis, G.; Filippatos, T.; Elisaf, M. Electrolyte disorders associated with the use of anticancer drugs. Eur. J. Pharmacol. **2016**, *777*, 78-87, doi:10.1016/J.EJPHAR.2016.02.064.

[11] Verzicco, I.; Regolisti, G.; Quaini, F.; Bocchi, P.; Brusasco, I.; Ferrari, M.; Passeri, G.; Cannone, V.; Coghi, P.; Fiaccadori, E.; et al. Electrolyte Disorders Induced by Antineoplastic Drugs. Front. Oncol. **2020**, *10*, doi:10.3389/FONC.2020.00779.

[12] Bendelow, J.; Apps, E.; Jones, L.; Poston, G. Carcinoid syndrome. Eur. J. Surg. Oncol. **2008**, *34*, 289-296, doi:10.1016/J.EJSO.2007.07.202.

[13] Ikuta, S.; Yasui, C.; Kawanaka, M.; Aihara, T.; Yoshie, H.; Yanagi, H.; Mitsunobu, M.; Sugihara, A.; Yamanaka, N. Watery diarrhea, hypokalemia and achlorhydria syndrome due to an adrenal pheochromocytoma. World J. Gastroenterol. **2007**, *13*, 4649-4652, doi:10.3748/WJG.V13.I34.4649.

[14] Torpy, D.; Mullen, N.; Ilias, I.; Nieman, L. Association of hypertension and hypokalemia with Cushing's syndrome caused by ectopic ACTH

secretion: a series of 58 cases. Ann. N. Y. Acad. Sci. **2002**, *970*, 134-144, doi:10.1111/J.1749-6632.2002.TB04419.X.

[15] Launay-Vacher, V.; Oudard, S.; Janus, N.; Gligorov, J.; Pourrat, X.; Rixe, O.; Morere, J.; Beuzeboc, P.; Deray, G. Prevalence of Renal Insufficiency in cancer patients and implications for anticancer drug management: the renal insufficiency and anticancer medications (IRMA) study. Cancer **2007**, *110*, 1376-1384, doi:10.1002/CNCR.22904.

[16] Born-Frontsberg, E.; Reincke, M.; Rump, L.; Hahner, S.; Diederich, S.; Lorenz, R.; Allolio, B.; Seufert, J.; Schirpenbach, C.; Beuschlein, F.; et al. Cardiovascular and cerebrovascular comorbidities of hypokalemic and normokalemic primary aldosteronism: results of the German Conn's Registry. J. Clin. Endocrinol. Metab. **2009**, *94*, 1125-1130, doi:10.1210/JC.2008-2116.

[17] Nagasaki, A.; Takamine, W.; Takasu, N. Severe hyperkalemia associated with "alternative" nutritional cancer therapy. Clin. Nutr. **2005**, *24*, 864-865, doi:10.1016/J.CLNU.2005.06.009.

[18] te Dorsthorst, R.P.M.; Hendrikse, J.; Vervoorn, M.T.; van Weperen, V.Y.H.; van der Heyden, M.A.G. Review of case reports on hyperkalemia induced by dietary intake: not restricted to chronic kidney disease patients. *Eur. J. Clin. Nutr. 2018 731* **2018**, *73*, 38-45, doi:10.1038/s41430-018-0154-6.

[19] Gupta, A.; Moore, J.A. Tumor Lysis Syndrome. *JAMA Oncol.* **2018**, *4*, 895-895, doi:10.1001/JAMAONCOL.2018.0613.

[20] Cairo, M.; Coiffier, B.; Reiter, A.; Younes, A. Recommendations for the evaluation of risk and prophylaxis of tumour lysis syndrome (TLS) in adults and children with malignant diseases: an expert TLS panel consensus. Br. J. Haematol. **2010**, *149*, 578-586, doi:10.1111/J.1365-2141.2010.08143.X.

[21] Mirrakhimov, A.E.; Ali, A.M.; Khan, M.; Barbaryan, A. Tumor Lysis Syndrome in Solid Tumors: An up to Date Review of the Literature. Rare Tumors **2014**, *6*, 68-75, doi:10.4081/RT.2014.5389.

[22] Lam, K.; Lo, C. Metastatic tumours of the adrenal glands: a 30-year experience in a teaching hospital. Clin. Endocrinol. (Oxf). **2002**, *56*, 95-101, doi:10.1046/J.0300-0664.2001.01435.X.

[23] Schulz, P.; Prüss, H. Hirnsymptome bei Carcinomatose "-Hermann Oppenheim and an Early Description of a Paraneoplastic Neurological Syndrome. J. Hist. Neurosci. **2015**, *24*, 371-377, doi:10.1080/0964704X.2015.1021120.

[24] Pelosof, L.C.; Gerber, D.E. Paraneoplastic Syndromes: An Approach to Diagnosis and Treatment. Mayo Clin. Proc. **2010**, *85*, 838, doi:10.4065/MCP.2010.0099.

[25] Fan, L.; Zhuang, Y.; Wang, Y.; Liu, X.; Liu, D.; Xiang, B.; He, M.; Zhang, Z.; Li, Y.; Wang, Y.; et al. Association of hypokalemia with cortisol and ACTH levels in Cushing's disease. Ann. N. Y. Acad. Sci. **2020**, *1463*, 60-66, doi:10.1111/NYAS.14205.

[26] Barbosa, S.; Rodien, P.; Leboulleux, S.; Niccoli-Sire, P.; Kraimps, J.; Caron, P.; Archambeaud-Mouveroux, F.; Conte-Devolx, B.; Rohmer, V. Ectopic adrenocorticotropic hormone-syndrome in medullary carcinoma of the thyroid: a retrospective analysis and review of the literature. Thyroid **2005**, *15*, 618-623, doi:10.1089/THY.2005.15.618.

[27] Teves, D. Clinical approach of Cushing syndrome resulting from

ACTH-producing metastatic neuroendocrine tumor. Endocrinologist **2005**, *15*, 401-404.

[28] Morandi, U.; Casali, C.; Rossi, G. Bronchial typical carcinoid tumors. Semin. Thorac. Cardiovasc. Surg. **2006**, *18*, 191-198, doi:10.1053/J. SEMTCVS.2006.08.005.

[29] Nimalasena, S.; Freeman, A.; Harland, S. Paraneoplastic Cushing's syndrome in prostate cancer: a difficult management problem. BJU Int. **2008**, *101*, 424-427, doi:10.1111/ J.1464-410X.2007.07294.X.

[30] D'Amico, M.; Gasparoli, L.; Arcangeli, A. Potassium Channels: Novel Emerging Biomarkers and Targets for Therapy in Cancer. Recent Pat. Anticancer. Drug Discov. **2013**, *8*, 53-65, doi:10.2174/1574892811308010053.

[31] Lastraioli, E.; Iorio, J.; Arcangeli, A. Ion channel expression as promising cancer biomarker. Biochim. Biophys. Acta - Biomembr. **2015**, *1848*, doi:10.1016/j.bbamem.2014.12.016.

[32] Hemmerlein, B.; Weseloh, R.M.; Mello de Queiroz, F.; Knötgen, H.; Sánchez, A.; Rubio, M.E.; Martin, S.; Schliephacke, T.; Jenke, M.; Heinz-Joachim-Radzun; et al. Overexpression of Eag1 potassium channels in clinical tumours. *Mol. Cancer 2006 51* **2006**, *5*, 1-13, doi:10.1186/1476-4598-5-41.

[33] Ousingsawat, J.; Spitzner, M.; Puntheeranurak, S.; Terracciano, L.; Tornillo, L.; Bubendorf, L.; Kunzelmann, K.; Schreiber, R. Expression of Voltage-Gated Potassium Channels in Human and Mouse Colonic Carcinoma. Clin. Cancer Res. **2007**, *13*, 824-831, doi:10.1158/1078-0432.CCR-06-1940.

[34] Abdul, M.; Hoosein, N. Voltage-gated potassium ion channels in colon cancer. Oncol. Rep. **2002**, *9*, 961-964, doi:10.3892/OR.9.5.961.

[35] Lastraioli, E.; Guasti, L.; Crociani, O.; Polvani, S.; Hofmann, G.; Witchel, H.; Bencini, L.; Calistri, M.; Messerini, L.; Scatizzi, M.; et al. herg1 Gene and HERG1 Protein Are Overexpressed in Colorectal Cancers and Regulate Cell Invasion of Tumor Cells. *Cancer Res.* **2004**, *64*, doi:10.1158/0008-5472. CAN-03-2360.

[36] Kim, C.J.; Cho, Y.G.; Jeong, S.W.; Kim, Y.S.; Kim, S.Y.; Nam, S.W.; Lee, S.H.; Yoo, N.J.; Lee, J.Y.; Park, W.S. Altered expression of KCNK9 in colorectal cancers. APMIS **2004**, *112*, 588-594, doi:10.1111/J.1600-0463.2004. APM1120905.X.

[37] Oeggerli, M.; Tian, Y.; Ruiz, C.; Wijker, B.; Sauter, G.; Obermann, E.; Güth, U.; Zlobec, I.; Sausbier, M.; Kunzelmann, K.; et al. Role of KCNMA1 in breast cancer. PLoS One **2012**, *7*, doi:10.1371/JOURNAL.PONE.0041664.

[38] Khaitan, D.; Sankpal, U.; Weksler, B.; Meister, E.; Romero, I.; Couraud, P.; Ningaraj, N. Role of KCNMA1 gene in breast cancer invasion and metastasis to brain. BMC Cancer **2009**, *9*, doi:10.1186/1471-2407-9-258.

[39] Bloch, M.; Ousingsawat, J.; Simon, R.; Schraml, P.; Gasser, T.; Mihatsch, M.; Kunzelmann, K.; Bubendorf, L. KCNMA1 gene amplification promotes tumor cell proliferation in human prostate cancer. Oncogene **2007**, *26*, 2525-2534, doi:10.1038/SJ.ONC. 1210036.

[40] Altintas, D.; Allioli, N.; Decaussin, M.; de Bernard, S.; Ruffion, A.; Samarut, J.; Vlaeminck-Guillem, V. Differentially expressed androgen-regulated genes in androgen-sensitive tissues reveal potential biomarkers of early prostate

cancer. PLoS One **2013**, *8*, doi:10.1371/JOURNAL.PONE.0066278.

[41] Brevet, M.; Ahidouch, A.; Sevestre, H.; Merviel, P.; El Hiani, Y.; Robbe, M.; Ouadid-Ahidouch, H. Expression of K+ channels in normal and cancerous human breast. *Histol. Histopathol.* **2008**, *23*, 965-972, doi:10.14670/HH-23.965.

[42] Haren, N.; Khorsi, H.; Faouzi, M.; Ahidouch, A.; Sevestre, H.; Ouadid-Ahidouch, H. Intermediate conductance Ca2+ activated K+ channels are expressed and functional in breast adenocarcinomas: correlation with tumour grade and metastasis status. *Histol. Histopathol.* **2010**, *25*, 1247-1255, doi:10.14670/HH-25.1247.

[43] Bulk, E.; Ay, A.; Hammadi, M.; Ouadid-Ahidouch, H.; Schelhaas, S.; Hascher, A.; Rohde, C.; Thoennissen, N.; Wiewrodt, R.; Schmidt, E.; et al. Epigenetic dysregulation of KCa 3.1 channels induces poor prognosis in lung cancer. *Int. J. cancer* **2015**, *137*, 1306-1317, doi:10.1002/IJC.29490.

[44] Lai, W.; Liu, L.; Zeng, Y.; Wu, H.; Xu, H.; Chen, S.; Chu, Z. KCNN4 Channels participate in the EMT induced by PRL-3 in colorectal cancer. *Med. Oncol. 2013 302* **2013**, *30*, 1-8, doi:10.1007/S12032-013-0566-Z.

[45] Bonito, B.; Sauter, D.; Schwab, A.; Djamgoz, M.; Novak, I. K Ca 3.1 (IK) modulates pancreatic cancer cell migration, invasion and proliferation: anomalous effects on TRAM-34. Pflugers Arch. **2016**, *468*, 1865-1875, doi:10.1007/S00424-016-1891-9.

[46] Ohya, S.; Kimura, K.; Niwa, S.; Ohno, A.; Kojima, Y.; Sasaki, S.; Kohri, K.; Imaizumi, Y. Malignancy grade-dependent expression of K+-channel subtypes in human prostate cancer. J. Pharmacol. Sci. **2009**, *109*, 148-151, doi:10.1254/JPHS.08208SC.

[47] Abdul, M.; Hoosein, N. Reduced Kv1.3 potassium channel expression in human prostate cancer. *J. Membr. Biol.* **2006**, *214*, 99-102, doi:10.1007/S00232-006-0065-7.

[48] Brevet, M.; Fucks, D.; Chatelain, D.; Regimbeau, J.; Delcenserie, R.; Sevestre, H.; Ouadid-Ahidouch, H. Deregulation of 2 potassium channels in pancreas adenocarcinomas: implication of KV1.3 gene promoter methylation. Pancreas **2009**, *38*, 649-654, doi:10.1097/MPA.0B013E3181A56EBF.

[49] Girault, A.; Privé, A.; Trinh, N.; Bardou, O.; Ferraro, P.; Joubert, P.; Bertrand, R.; Brochiero, E. Identification of KvLQT1 K+ channels as new regulators of non-small cell lung cancer cell proliferation and migration. Int. J. Oncol. **2014**, *44*, 838-848, doi:10.3892/IJO.2013.2228.

[50] Ding, X.; Wang, X.; Luo, H.; Tan, S.; Gao, S.; Luo, B.; Jiang, H. Expression and prognostic roles of Eag1 in resected esophageal squamous cell carcinomas. Dig. Dis. Sci. **2008**, *53*, 2039-2044, doi:10.1007/S10620-007-0116-7.

[51] Ding, X.; Yan, J.; An, P.; Lü, P.; Luo, H. Aberrant expression of ether à go-go potassium channel in colorectal cancer patients and cell lines. World J. Gastroenterol. **2007**, *13*, 1257-1261, doi:10.3748/WJG.V13.I8.1257.

[52] Ding, X.; Luo, H.; Jin, X.; Yan, J.; Ai, Y. Aberrant expression of Eag1 potassium channels in gastric cancer patients and cell lines. Med. Oncol. **2007**, *24*, 345-350, doi:10.1007/S12032-007-0015-Y.

[53] Ding, X.; Luo, H.; Luo, B.; Xu, D.; Gao, S. Overexpression of hERG1 in resected esophageal squamous cell carcinomas: a marker for poor prognosis. J. Surg. Oncol. **2008**, *97*, 57-62, doi:10.1002/JSO.20891.

[54] Lastraioli, E.; Taddei, A.; Messerini, L.; Comin, C.E.; Festini, M.; Giannelli, M.; Tomezzoli, A.; Paglierani, M.; Mugnai, G.; De Manzoni, G.; et al. hERG1 channels in human esophagus: Evidence for their aberrant expression in the malignant progression of Barrett's esophagus. J. Cell. Physiol. **2006**, *209*, doi:10.1002/jcp.20748.

[55] Lastraioli, E.; Lottini, T.; Iorio, J.; Freschi, G.; Fazi, M.; Duranti, C.; Carraresi, L.; Messerini, L.; Taddei, A.; Ringressi, M.N.; et al. hERG1 behaves as biomarker of progression to adenocarcinoma in Barrett's esophagus and can be exploited for a novel endoscopic surveillance. *Oncotarget* **2016**, *7*, doi:10.18632/oncotarget.11149.

[56] Ding, X.; Yang, W.; Gao, S.; Wang, W.; Li, Z.; Hu, W.; Li, J.; Luo, H. Prognostic significance of hERG1 expression in gastric cancer. Dig. Dis. Sci. **2010**, *55*, 1004-1010, doi:10.1007/S10620-009-0834-0.

[57] Crociani, O.; Lastraioli, E.; Boni, L.; Pillozzi, S.; Romoli, M.R.; D'Amico, M.; Stefanini, M.; Crescioli, S.; Taddei, A.; Bencini, L.; et al. HERG1 channels regulate VEGF-A secretion in human gastric cancer: Clinicopathological correlations and therapeutical implications. Clin. Cancer Res. **2014**, *20*, doi:10.1158/1078-0432.CCR-13-2633.

[58] Lastraioli, E.; Bencini, L.; Bianchini, E.; Romoli, M.R.; Crociani, O.; Giommoni, E.; Messerini, L.; Gasperoni, S.; Moretti, R.; di Costanzo, F.; et al. hERG1 channels and Glut-1 as independent prognostic indicators of worse outcome in stage I and II colorectal cancer: A pilot study. Transl. Oncol. **2012**, *5*, doi:10.1593/tlo.11250.

[59] Iorio, J.; Lastraioli, E.; Tofani, L.; Petroni, G.; Antonuzzo, L.; Messerini, L.; Perrone, G.; Caputo, D.; Francesconi, M.; Amato, M.M.; et al. hERG1 and HIF-2α Behave as Biomarkers of Positive Response to Bevacizumab in Metastatic Colorectal Cancer Patients. Transl. Oncol. **2020**, *13*, doi:10.1016/j.tranon.2020.01.001.

[60] Lastraioli, E.; Perrone, G.; Sette, A.; Fiore, A.; Crociani, O.; Manoli, S.; D'Amico, M.; Masselli, M.; Iorio, J.; Callea, M.; et al. HERG1 channels drive tumour malignancy and may serve as prognostic factor in pancreatic ductal adenocarcinoma. Br. J. Cancer **2015**, *112*, doi:10.1038/bjc.2015.28.

[61] Stringer, B.K.; Cooper, A.G.; Shepard, S.B. Overexpression of the G-Protein Inwardly Rectifying Potassium Channel 1 (GIRK1) in Primary Breast Carcinomas Correlates with Axillary Lymph Node Metastasis. Cancer Res. **2001**, *61*.

[62] Williams, S.; Bateman, A.; O'Kelly, I. Altered Expression of Two-Pore Domain Potassium (K2P) Channels in Cancer. PLoS One **2013**, *8*, e74589, doi:10.1371/JOURNAL.PONE.0074589.

[63] Hou, X.; Tang, L.; Li, X.; Xiong, F.; Mo, Y.; Jiang, X.; Deng, X.; Peng, M.; Wu, P.; Zhao, M.; et al. Potassium Channel Protein KCNK6 Promotes Breast Cancer Cell Proliferation, Invasion, and Migration. *Front. cell Dev. Biol.* **2021**, *9*, doi:10.3389/FCELL.2021.616784.

[64] Mu, D.; Chen, L.; Zhang, X.; See, L.-H.; Koch, C.M.; Yen, C.; Tong, J.J.; Spiegel, L.; Nguyen, K.C..; Servoss, A.; et al. Genomic amplification and oncogenic properties of the KCNK9 potassium channel gene. Cancer Cell **2003**, *3*, 297-302, doi:10.1016/S1535-6108(03)00054-0.

[65] Alvarez-Baron, C.P.; Jonsson, P.; Thomas, C.; Dryer, S.E.; Williams, C. The Two-Pore Domain Potassium

Channel KCNK5: Induction by Estrogen Receptor α and Role in Proliferation of Breast Cancer Cells. Mol. Endocrinol. **2011**, *25*, 1326-1336, doi:10.1210/ME.2011-0045.

[66] Voloshyna, I.; Besana, A.; Castillo, M.; Matos, T.; Weinstein, I.B.; Mansukhani, M.; Robinson, R.B.; Cordon-Cardo, C.; Feinmark, S.J. TREK-1 Is a Novel Molecular Target in Prostate Cancer. Cancer Res. **2008**, *68*, 1197-1203, doi:10.1158/0008-5472.CAN-07-5163.